UNCERTAINTY IN THE ELECTRIC POWER INDUSTRY

* A list of the early publications in the series is at the end of the book *

UNCERTAINTY IN THE ELECTRIC POWER INDUSTRY
Methods and Models for Decision Support

CHRISTOPH WEBER

University of Stuttgart, *Institute for Energy Economics and Rational of Use of Energy*

 Springer

Christoph Weber
Institut for Energy, Economics and Rational Use of Energy
University of Stuttgart, Germany

Library of Congress Cataloging-in-Publication Data

Weber, Christoph.
    Uncertainty in the electric power industry : methods and models for decision support /
Christoph Weber.
        p. cm. – (International series in operations research & management science; 77)
    Includes bibliographical research and index.
    ISBN 0-387-23047-5 – ISBN 0-387-23048-3 (e-book)
    1. Electric utilities. 2. Electric utilities—Deregulation. 3. Electric utilities—Planning. I.
Title. II. Series.

HD9685.A2W43 2004
 333.793'23—dc22

                                                              2004057600

Printed in the United States of America.

9 8 7 6 5 4 3 2 1   SPIN 11052869

springeronline.com

For Dorothee,
Simon and Miriam

Contents

List of Figures

xiv

List of Tables

Preface

Liberalization and privatization in the electricity industry have lead to increased competition among utilities. At the same time, utilities are now exposed more than ever to risk and uncertainties, which they cannot pass on to their customers through price increases as in a regulated environment. Especially electricity generating companies have to face volatile wholesale prices, fuel price uncertainty, limited long-term hedging possibilities and huge, to a large extent sunk investments. In this context, the present book aims at an integrative view on the decision problems that power companies have to tackle.

The central challenge is thereby the optimization of generation and trading portfolios under uncertainty - and by purpose this is also a central chapter in the book. But this optimization is not possible without a profound understanding and detailed models of market and price developments as well as of competitors' behavior. For market and price modeling the focus is thereby on an innovative integrative approach, which combines fundamental and finance-type models.

The optimization of the portfolios has furthermore to go along with an adequate management of the corresponding risks. Here the concept of Integral-Earnings-at-Risk is worked out, which takes into account market structure and market liquidity. It provides a theoretically justified alternative going beyond a simplistic combination of Value-at-Risk and Profit-at-Risk measures.

After more than one decade of liberalization efforts, also the longer-term investment decisions are becoming more pressing for the electricity generation companies. These require both an assessment of future technology trends and of long-term price and capacity developments.

Especially fuel cells are a key challenge to the power industry and therefore adequate methods for technology assessment are developed and applied to this example. Another key issue is the development of optimal investment strategies under fuel price uncertainty. This requires models of investment under uncertainty, which combine the real options approach with models of endogenous market price equilibria as developed within peak-load pricing theory.

The primary intention of this book is not to provide an in-depth discussion of the regulatory challenges at hand after more than one decade of electricity market deregulation around the world – but analyzing the key decision problems of players in the industry certainly is a useful and necessary first step when aiming at the design of efficient and robust electric power markets. And by bringing together material and approaches from different disciplines, the volume at hand hopefully helps both practitioners and academics to identify the adequate models for the challenges they have to cope with.

Stuttgart, July 2004   Christoph Weber

Acknowledgments

This work would not have been possible without the support by many colleagues and friends. First I would like to express my gratitude to Prof. Alfred Voß, who gave me the opportunity to do the research underlying this book at IER, University of Stuttgart and who was the main supervisor, when this work was submitted as "Habilitationsschrift" at Stuttgart University. His interest in new methods and his profound knowledge in system analysis have motivated many of the approaches presented here. My thanks go furthermore to Prof. Klaus Hein and Prof. Erich Zahn, who supported this thesis as co-supervisors and who with their respective competences have been a strong encouragement for me to devote as much attention to technical details as to economic thinking.

I am also deeply indebted to all colleagues, with whom I collaborated within the research group on Energy Management and Energy Use (EAM) at IER and beyond. Without their support and the constructive and inspiring research climate, this work could not have been completed. Specifically, I would like to thank Dr. Kai Sander, Heike Brand, Christopher Hoeck, Derk Swider, Dr. Eva Thorin, Dr. Markus Blesl, Dr. Nikolaus Kramer, Henrik Specht, Ingo Ellersdorfer, Alfred Hoffmann and Dr. Andreas Schuler for helping me with data and models. Without them and the many fruitful discussions we had together the work would lack if not its essence then many of its flavors. My special thanks go to my father, Dr. Fritz Weber, for proofreading the manuscript and looking at it as a non-specialist.

But without the patience and the love of my wife and children, their acceptance of many hour and day spent on this and other work, this book would never have seen the daylight.

Chapter 1

INTRODUCTION

Starting in Chile, the UK and Norway, liberalization and privatization have been a major theme in the electricity industry during the last decade. Through the introduction of competition and economic considerations, governments around the world have attempted to obtain more reliable and cheaper services for the electricity customers. A major step in Europe has been the directive of the European Commission in the end of 1996 (EU 1997), requiring the stepwise opening of electricity markets in the European Union, ending with a fully competitive market at the latest in 2010. Also in the US, many federal states have taken steps towards competitive and liberalized markets, with California and several East coast states being among the first movers. But California is nowadays often cited as the pre-eminent example for the risks and difficulties associated with liberalization. Adding to this the Enron collapse at the end of 2001, the strive for liberalization has considerably been slowed down and the uncertainties and risks inherent in liberalized electricity markets are much more in view.

In order to avoid throwing away the baby with the bath, one has to look carefully at the decision situation faced by the different actors in liberalized electricity markets. Special attention has thereby to be devoted to the generation companies, since they are exposed to the risks of competition and at the same time have to afford huge and to a large extent irreversible investments. The uncertainties, which these companies are facing in the new electricity market, include notably the future development of:

- product prices for electricity,
- world market prices for primary energy carriers (coal, gas and oil),
- technology (e.g. distributed generation),
- regulation and political context (including environmental policy),
- behavior of competitors,
- availability of plants,

- demand growth.

These factors have to be accounted for both in operative decision making and in strategic planning and require an increased use of mathematical models for decision support.

In this context, the present study aims to develop models for supporting the energy management in the electricity industry. Thereby various methods developed in operations research are employed and combined to provide models of practical relevance and applicability. A particular focus is on the concepts of stochastic processes and stochastic optimization, but also game and control theory approaches and finance concepts like value-at-risk or profit-at-risk are employed. At the same time, the analysis emphasizes the need to include sufficient technical detail to account for the specificities of the electricity industry, notably the non-storability of electricity and the grid dependency.

In the following, the basis for the subsequent analyses is first laid through a review of the current situation in various countries and the relevant market structures in chapter 2 and a brief recapitulation of the basics from decision sciences and mathematics in chapter 3. Then models to cope with the key uncertainty of price developments are discussed in chapter 4. Chapter 5 is devoted to modeling the interactions between the different players on the market. The operative decisions of unit commitment, dispatch and (short and medium term) portfolio choice are discussed in chapter 6. The controlling and management of the associated risks is then covered in chapter 7. Longer-term aspects are analyzed in chapter 8 and 9. Thereby, chapter 8 focuses on the role of uncertain technology developments and chapter 9 analyses the optimal investment decisions in this context.

Chapter 2

DEREGULATION AND MARKETS IN THE ELECTRICITY INDUSTRY

Traditionally the electricity business used to be organized along the physical energy flow from electricity generation through the transmission and distribution grid to the final customer (cf. Figure 2-1). Often the whole chain was vertically integrated into one company (e.g. EDF in France) or at least the generation and transmission business was integrated (e.g. former Preussenelektra in Germany). Many of these utilities were also fully or partly state-owned.

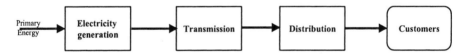

Figure 2-1. Structure of the regulated electricity industry

Deregulation, which has been a major theme world-wide in the electricity industry during the last decade, has brought competition and privatization for many parts of the formerly heavily regulated electricity industry. Regulation had especially been justified by the natural monopolies which arise due to the network dependency of electricity production and consumption. And deregulation has taken as its starting point the observation that certainly the transmission and distribution of electricity give rise to natural monopolies (in formal terms: subadditive cost functions, cf. Baumol et al. 1982) but that this is not necessarily true for the generation and retail sales businesses in the industry (cf. e.g. Gilsdorf 1995, Pineau 2002). Competition has therefore been introduced at the wholesale and the retail level. In parallel to the physical load flow, market places and actors acting on these market places have emerged as indicated in Figure 2-2. But this figure gives only the general lines. To what extent the different elements are in place in the different OECD countries is discussed in the following section. Then, a closer look is taken at the market structures put in place at the wholesale level, focusing on the situation in continental Europe and notably in Germany.

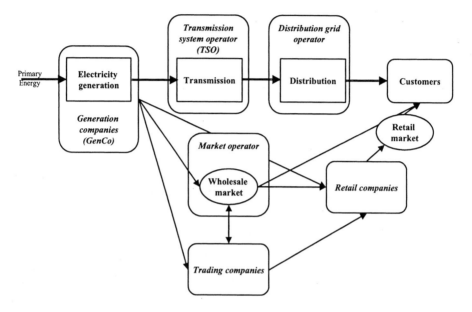

Figure 2-2. Structure of the deregulated electricity industry

1.   STATUS OF DEREGULATION

Many countries have so far focused on the establishment of wholesale competition (cf. Table 3-1). This requires notably the unbundling of transmission and generation activities and a non-discriminatory access to the transmission infrastructure for all generation companies. Since generation is viewed as a competitive market, there are at first sight few reasons for public ownership and deregulation may be accompanied by privatization.[1] The benefits and drawbacks of alternative ownership and operation models

---

[1] The arguments in favor of privatization stemming from various theories are reviewed by Pollitt (1997). Privatization has notably been an important part of the restructuring policies in the UK and Chile, two of the early movers in the field of electricity industry deregulation. On the other hand, Norway still mostly has state and municipally owned generators. And Bergman et al. (2000) make the point that it is less the ownership character which matters but the concentration of ownership.

Table 2-1. Wholesale market and transmission system structures in OECD countries in 2003

Country	Wholesale market			Transmission system operators (TSO)		
	Year of intro-duction	Type	Market places	number	owner-ship	Unbundling with generation companies
EU						
Austria	2002	Volunt.	EXAA	3	Mixed	Legal separation
Belgium	Only bilateral trade			1 (Elia)	Mixed	Legal separation
Denmark	1999/2000	Volunt.	Nordpool	2	Public	Legal separation
Finland	1997	Volunt.	Nordpool/ EL_EX	1 (Fingrid)	Public	Ownership sep.
France	2001	Volunt.	Powernext	1	Public	Accounting sep.
Germany	2000	Volunt.	EEX	4	Mixed	Accounting sep.
Greece	2002	Volunt.	HTSO	1 (HTSO)	Mixed	Accounting sep.
Ireland	Planned	Volunt.	ESBNat.Grid /EIRGRID	1 (ESB Nat. Grid)	Public	Legal separation
Italy	Planned	Volunt.	GME	1 (GRTN)	Public	Operational sep.
Luxembourg	Only bilateral trade			1 (CEGEDEL)	Mixed	Legal separation
Netherlands	1999	Volunt.	APX	1 (Tennet)	Public	Legal separation
Portugal	2001	Volunt.	REN	1 (REN)	Public	Legal separation
Spain	1998	Volunt.	OMEL	1 (REE)	Private	Ownership sep.
Sweden	1996	Volunt.	Nordpool	1 (Svenska Kraftnät)	Public	Ownership sep.
United Kingdom *incl.:*	Differences between regions			3	Public	Ownership sep.
England & Wales	1990	Volunt.	UKPX and APX UK	1	Public	Ownership sep.
Rest of Europe						
Norway	1993	Volunt.	Nordpool	1 (StatNet)	Public	Ownership sep.
Switzerland	Open (popular vote against liberalization in 2003))			1 (ETRANS)	Mixed	No separation required
Turkey	Planned			1(TEIAS)	Public	Legal separation
North America						
Canada	Differences between regions: Ontario and Alberta have introduced full competition, the other 8 provinces only partly at wholesale level					Depending on province
US, *incl.:*	Differences between states, 23 states have deregulated			> 140 control areas	Mixed	Depending on state
California	1998	Volunt	CalPX (until 2000)	1	Public	Ownership sep.
Pennsylvania, New Jersey, Maryland	1997	Volunt.	PJM	1	Mixed	Operation separation
Asia and Oceania						
Japan	Planned			10	Mixed	Planned
South Korea	2001	Volunt.	Korea Power Exchange	1	Public	Legal sep., owner-ship sep. planned
Australia	1998	Mandat.	NEM	9	Mixed	Operation sep.
New Zealand	1996	Volunt.	NZEM	1 (Transpower)	Public	Ownership sep.

Sources: IEA (2001), Madlener, Kaufmann (2002), Prospex (2003), EIA (2003), Internet sites of TSOs and market places

for the transmission grid are on the other hand more controversially discussed[2]. In practice, mostly solutions have been chosen which did not require forced divestiture, since there are often strong legislative and political barriers to such steps (e.g. in Germany and California). But as Bergman et al. (2000) emphasize, accounting and even legal separation between transmission and generation entities may not be sufficient to ensure discrimination free network access[3].

Competition at the retail level has been introduced in most countries much later than wholesale competition and even today does often not cover all customer segments (cf. Table 2-2). This is at first sight surprising given that the overall aim of deregulation is to increase public welfare not the last through lower consumer prices. But several practicalities are put forward against extending the competition to the retail level. Major points which arise in the debate[4] include the following:

Efforts needed for unbundling distribution and retail services (sometimes also called supply services) are higher than for unbundling generation and transmission, given that often many local and regional distribution companies exist, compared to a few transmission operators (cf. Table 2-2).

Efforts necessary for metering, data transmission and billing are considerably higher in competitive retail markets. Ideally, all customers should be equipped with quarter-hourly or hourly meters – the use of profiling for determining approximate load shapes is an alternative used for small customers in many countries, but this leads also to considerable problems[5].

The two preceding points may lead to the conclusion that *additional transaction costs* through unbundled distribution and retail could be higher than the value added created in the pure retail business and thus overall costs for consumers are possibly higher in competitive retail markets than in monopolistic ones.

Electricity is nowadays such a basic good in industrialized countries that there is (or seems to be) a *public service obligation for electricity supply.*

---

[2]   The different business models currently envisaged in the U.S. are discussed by Oren et al. (2002), Sharma (2002).

[3]   Brunekreeft (2002) provides a detailed analysis of the German situation and emphasizes that the principle of a level playing field is violated under the current regulation. Yet he points at the possibility that the current procedures of the Bundeskartellamt under the general anti-trust law may be sufficient to prevent competition distortion.

[4]   The current on-going debate in the U.S. is summarized in HEPG(2002)

[5]   Cf. notably the debate in Germany on the use of analytical vs. synthetic load profiles (e.g. Pohlmann, Pospischill 1999). Also the price signals to consumers (and distributed generators) are necessarily inadequate, if no actual metering is done.

Table 2-2. Retail market and distribution grid structures in OECD countries in 2003

Country	Retail competition		Distribution grid operator			Unbundling of
	Year of intro-duction	Degree of market opening Feb. 2003	Number	ownership	Unbundling with retail companies	generation and retail
EU						
Austria	1999	100 %	12 + small	Mixed	Accounting sep.	Accounting sep.
Belgium	2000	52 %	≈ 60	Mixed	Accounting sep.	Accounting sep.
Denmark	1998	100 %	≈ 50	Mixed	Legal separation	Legal separation
Finland	1995	100 %	≈ 100	Mixed	Accounting sep.	Accounting sep.
France	2000	37 %	1 + ≈ 170 small	Public	Accounting sep.	Accounting sep.
Germany	1998	100 %	≈ 600	Mixed	Accounting sep.	Accounting sep.
Greece	2001	34 %	1	Public	Accounting sep.	Accounting sep.
Ireland	2000	56 %	1	Public	Accounting sep.	Accounting sep.
Italy	2000	70%	1	Public	Operational sep.	Operational sep.
Luxemburg	1999	57 %	10	Mixed	Accounting sep.	Accounting sep.
Netherlands	1999	63 %	≈ 20	Mixed	Legal separation	Legal separation
Portugal	1995	45 %	4 + ≈ 30 small	Public	Legal separation	Legal separation
Spain	1998	100 %	≈ 120	Public, to be privatized	Legal separation	Legal separation
Sweden	1996	100 %	8 + ≈ 250 small	Mostly public	Legal separation	Legal separation
UK	1996	100 %	15	Public	Legal separation	Legal separation
Rest of Europe						
Norway	1991	100 %	≈ 200	Mostly public	Legal separation	Legal separation
Switzerland	Not yet		≈ 900	Mixed	Not required	Not required
Turkey	2003	23 %	1	Public	Not required	Not required
North America						
Canada	Differences between regions: Ontario and Alberta have introduced retail competition, not the other 8 provinces					Depending on province
US, *incl.:*	Differences between states, 12 states have implemented far reaching deregulation including retail competition					Depending on state
California	1998 (suspended in 2001)	100 %	3 + ≈ 54 small	Mixed	Accounting sep.	Partly divestiture required
Pennsylvania	1999	100 %	≈ 10	Mixed	Accounting sep.	Accounting sep.
Asia and Oceania						
Japan	1999	30 %	10	Mixed	Not required	Not required
South Korea	Planned	-	1	Public, to be privatized	Planned	Planned
Australia	1998	Depends on state	10	Mixed	Accounting sep.	Ownership sep.
New Zealand	1994	100 %	29	Public	Accounting sep.	Ownership sep.

Sources: IEA (2001), CEC (2003), CEC (2001), EIA (2003), Joskow (2002), Internet sites of
grid operators and regulatory bodies

Actual percentages of customers switching to other suppliers are rather low in most markets where retail competition has already been introduced.

The last argument may well be countered by reminding that even if actually no supplier switch occurs, competition may be beneficial by obliging the former monopolist to offer more competitive prices[6]. The public service obligation may also be solved at least partly in competitive markets by transforming it into a connection obligation for the distribution grid company (which anyhow remains regulated) and by imposing principles of non-discrimination for grid charges and energy prices[7].

So the most pertinent arguments against introduction of competition at the retail level are in the view of the author those invoking the potential transaction costs arising with retail competition. An empirical analysis of these costs is certainly a valuable albeit tedious task, but is clearly beyond the scope of this book. The famous statement by Adam Smith "Consumption is the sole end and purpose of all production" (Smith 1962) may be invoked as a rather general, philosophical argument in favor of retail competition for all customer groups. More practical arguments will show up in the analysis of competition equilibria in chapter 5 and in chapter 7 on risk management.

2.   POWER AND RELATED MARKETS IN CONTINENTAL EUROPE

The key market for power deliveries in most countries is the day-ahead spot market. This trading is done in continental Europe mostly on an hourly basis (cf. Figure 2-3). In these countries, there is so far hardly any trading after the closure of the day-ahead market, albeit some tentatives for establishing an intraday market exist.

In the German Verbändevereinbarung (BDI et al. 2001) notably the possibility of resubmitting transmission schedules (which is a prerequisite of intraday trading) is foreseen in the case of power plant outages. But most deviations between scheduled and actual production and consumption are handled so far through the reserve power markets. According to the UCTE (Union of European Transmission Grid Operators) standards, reserve is divided into primary, secondary and tertiary (minute) reserve. The first two correspond to spinning reserves which can be activated within seconds

---

[6] This is the concept of contestable markets going back to Hayek (1937) and Machlup (1942), emphasized by Baumol et al. (1982) and applied e.g. to the gas industries by Knieps (2002)

[7] However a critical point is then who will provide the generation capacity used by the provider of last resort.

respectively within minutes, whereas the minute reserve may also include non-spinning reserve which can be brought on-line within a maximum delay of 15 minutes (notably gas turbines). For these reserves, specific markets have also been established - at least in Germany (cf. Swider, Weber 2003). There bids for primary and secondary reserve are submitted to each transmission grid operator on a bi-annual basis, whereas the bids for minute reserve are submitted on a day-to-day basis. The bids comprise usually a capacity and an energy price, with the capacity price describing the option value paid by the grid operator for the right to exercise the plant operation option. The energy price is on the contrary only paid if the option is exercised.

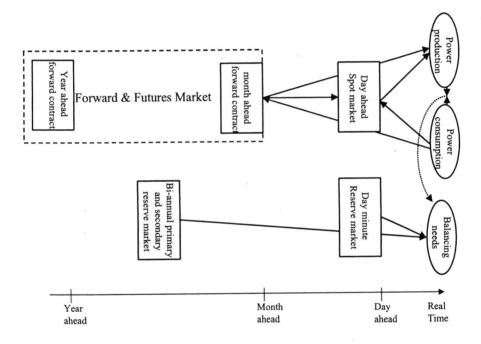

Figure 2-3. Trading structure in continental European markets

On the longer term derivative market, the market is also far from being fully developed. The shortest term liquid product is usually the month ahead forward (or future) contract and also the year-ahead product is traded rather frequently. All other products are much less liquid and even though quotes exist, it might be difficult to sell or buy larger quantities of these products without affecting the prices. So the principle of one continuous market with one uniform price, as put forward for example by Shuttleworth and Lieb-Doczy (2000), is far from being achieved in the continental European context. Besides historical and circumstantial reasons, two fundamental factors may explain, why no complete market structure with uniform prices emerges:

- The first is the persistent vertical integration of utilities, which leads, as discussed in chapter 5, to a limited liquidity on the power exchange markets
- The second is the dominance of thermal power plants in the generation park, for which the marginal generation cost is dependent on the time lag between unit commitment decision and operation and also on the operation mode in preceding and subsequent hours. So for the individual units no uniform marginal price exists and this makes notably a reserve market with combined capacity and energy bids more convenient.

Of course, this market structure has to be accounted for when developing practically relevant decision support models for the industry.

Chapter 3

DECISION MAKING AND UNCERTAINTIES IN THE ELECTRICITY INDUSTRY

In this chapter, the basis for the subsequent analyses is laid by first summarizing which are the key decision problems in the electricity industry (section 1). Then, the major uncertainties are discussed in section 2 and the formal framework used in the subsequent chapters is described in section 3.

1. DECISION PROBLEMS IN THE ELECTRICITY INDUSTRY

Decision problems in the electricity industry may be categorized along different lines. A first useful distinction is to look at the part of the value chain, where the decisions are taking place. Accordingly (cf. Figure 2-2), we may distinguish decisions taken by:
- Players in competitive segments
  - Generation companies
  - Trading companies
  - Retail companies
  - Holdings of vertically integrated companies
  - Electricity customers
- Players in regulated segments
  - Transmission companies and transmission system operators
  - Distribution companies and distribution system operators
- Regulators
  Another division is according to the impact of the decisions. Here the common distinction between

- Operational decisions: having a short-term impact and affecting only specific functional areas and
- Strategic decisions: having a long-term impact and/or affecting all functional areas within a firm

is useful.

Furthermore, decisions should also be distinguished according to the type of resources which they affect: human, financial and/or physical resources (material and production equipment). Of course, most decisions in a firm will affect in some way or another human and financial resources. But not all may directly have an effect on physical resources. For example, the selection of a candidate for a job opening has often no impact on the material and production equipment he or she will use. Also merger & acquisition decisions have not by themselves an impact on the physical production processes, albeit of course subsequent decisions may profoundly affect material flows and production equipment used.

These few categorizations – many more are possible of course – clearly illustrate the broad variety of decisions taken within the industry we are studying. And obviously any attempt to treat all of them in depth is condemned to fail. The focus of the following analyses is therefore on those decisions, which are at least to some extent specific to the electricity industry - i.e. those linked to the physical production processes and resources used. Furthermore, a restriction to those decisions which have to be rethought in the new, deregulated environment seems advisable.

Consequently, the following analyses will look in detail at those operative and strategic decisions in the generation and trading businesses which affect electricity generation. Transmission and distribution are not scrutinized, since these still remain monopolies. Nor decisions in electricity retailing are looked at, albeit those are now radically different from former times. But in that branch, methods and models already in use in other retail markets can be fruitfully applied[8]. Consequently, the focus is on models for *production scheduling, portfolio management and investment decisions* in the electricity industry.[9] Thereby not only models for optimal decision making will be derived, but also models to cope with the key uncertain factors affecting these decisions. The key interconnections between uncertain factors and decisions are illustrated in Figure 3-1.

---

[8] Examples of quantitative models and analyses in this field include Weber et al. (2001)

[9] Decisions on maintenance and retrofit are a further topic of considerable interest for electricity generation companies in liberalized markets. But here again, by and large similar methods to those in use in other sectors (e. g. chemical industry) may be applied.

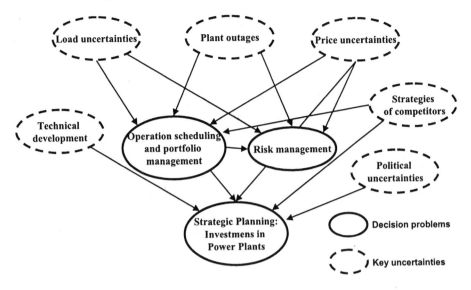

Figure 3-1. Uncertainties and key decisions in competitive electricity markets

The key uncertain factors will be scrutinized in the following section.

2. UNCERTAINTIES IN THE ELECTRICITY INDUSTRY

In order to develop good[10] models for decision making, it is essential to have a good understanding of the issues at stake. Therefore stylized facts about the key uncertainties mentioned in Figure 3-1 are presented in this section.

2.1 Market prices for primary energy carriers

Most of the electricity generated worldwide is produced from one of the following primary energy carriers (cf. e.g. EC 2002): coal, oil, gas, hydro and nuclear. For nuclear and hydro no public markets or trading platforms exist. In the case of nuclear fuels, this is certainly due both to concerns on proliferation of uranium for nuclear weapons and to the technological

---

[10] "Good" may for our purposes be provisionally defined as: fitting closely the observed reality while being at the same time as easy as possible to understand and to implement. Of course, more rigorous measures of model quality can be developed based on the formal concepts introduced in section 3.

complexity of the nuclear fuel chain, which requires long-term contracts to secure investments. Hydro energy (and energy from other renewables such as solar and wind) is not transportable and is therefore transformed on-site into electricity. Consequently, no national or even world-wide market place exists. A similar argument holds for lignite (brown coal), which is of considerable importance for the electricity generation in some countries, notably in Germany, where it contributes about 25 % to total generation. Lignite has a low specific calorific value compared to hard coal; therefore the transport of lignite would be comparatively very expensive. Consequently lignite is almost exclusively burnt in on-site power stations and hardly any national or international market exists.

For coal, oil and gas, selected market prices are shown in Figure 3-2. Market prices in other locations worldwide should normally not differ by more than the transportation costs from the prices shown here.

Figure 3-2 clearly shows that prices for oil are subject to considerable variations in the longer run. For example, crude oil prices have risen by a factor of three between the beginning of 1999 and the middle of 2000. But also other periods with rather low price variations exist. The average day to day variations (volatility) has been 2.3 % for the IPE products during the period described here.

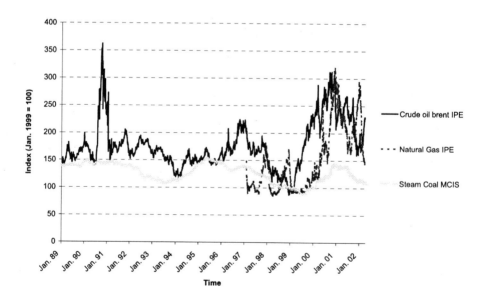

Figure 3-2. Market price indices for primary energy carriers

The price of coal has been more stable in comparison; it has hardly varied by more than a factor of 1.5 for the price index considered here. It is worth noting that coal is still mostly delivered to power plants based on long-term contracts and therefore for a long time period only monthly price quotes have been established. A major reason for the use of long-term contracts is that coal from different production sites can not be burnt in the same power plant without substantial technical modifications in the plants to adapt for variations in coal quality (e.g. sulfur and ash content, calorific value).

Natural gas has only become more recently a commodity traded on a day-to-day basis and even in many regions in the world (including continental Europe) it is not traded extensively so far. For a long time, the gas price used to be closely tied to the oil price through price formulas containing smoothing and lagging components. And even the recent British gas price developments shown in Figure 3-2 exhibit a clear lagged correlation of the prices. Another point worth noting is that natural gas prices exhibit considerable seasonal cycles. This is due to the limited storability of gas and the clear seasonal variations in gas demand, related to its use as heating fuel.

2.2   Product prices for electricity

In most electricity markets world-wide, spot market prices for electricity are characterized by (cf. also Johnson, Barz 1999):

- daily, weekly and seasonal cycles,
- high volatility,
- mean reversion and
- spikes.

This is illustrated for the case of the German market in Figure 3-3. Thereby daily price averages for peak and base products in the day-ahead market are shown, the hourly prices even exhibit higher volatility and spikes. Several analyses have highlighted that among all traded commodities, electricity exhibits the highest volatilities (e.g. Pilipovic 1998). A major reason for these strong fluctuations is the non-storability of electricity. It implies that at each moment in time supply has to match demand and in peak demand hours, prices may increase drastically, especially if some unforeseen plant outages lead to capacity shortages. It is worth mentioning that in hydro-dominated systems, such as notably the Nordic power market, electricity prices are much less subject to short-term fluctuations (cf. Figure 3-4). Even if electricity itself remains non-storable, the storability of water has as a consequence that electricity prices behave much more like those of a storable commodity.

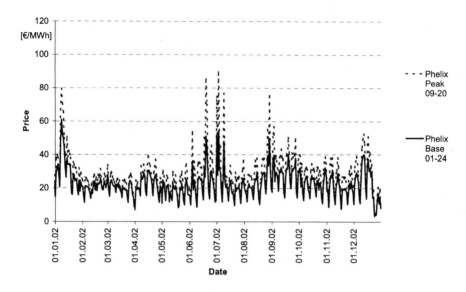

Source: EEX (2003)

Figure 3-3. Electricity spot market prices in Germany year 2002

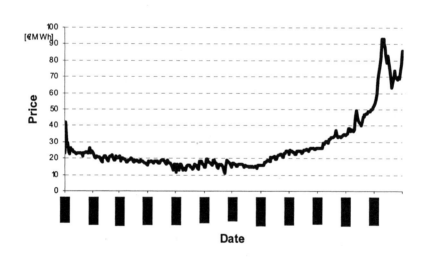

Source: Nordpool (2003)

Figure 3-4. Electricity spot market prices in the Nordpool area year 2002

Given the extreme price volatility of spot market prices, the establishment of electricity trading in the short term (such as on spot markets) has been in most markets soon followed by possibilities to sign up electricity contracts in advance. Although considerable differences exist between the products established in different markets, one may retain that in general future and forward prices for electricity contrarily to spot prices exhibit:

- no cycles,
- lower volatility,
- little mean reversion.

Figure 3-5 illustrates this for the case of the German market. The absence of cycles is of course due to the fact that the price developments of forwards or futures for one given expiry date are considered here. The forward curve, which includes forward prices for different monthly or quarterly delivery dates at a given moment in time, shows on the contrary seasonal patterns. The lower volatility of the future prices is related to the fact that the futures themselves are storable equities, since they may be purchased today and sold tomorrow or next month. This explains also why little mean-reversion in these prices is observed. As for any storable equity, arbitrage opportunities would arise if the (discounted and risk-adjusted) product price would not follow a martingale process, i.e. a stochastic process where the observed value today corresponds to the risk adjusted expected value for tomorrow (cf. e.g. Hull 2000).

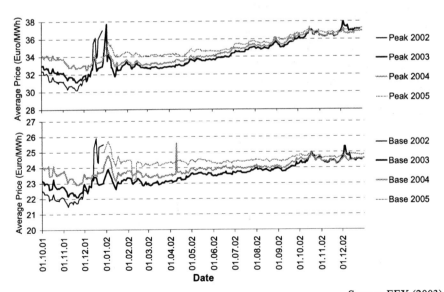

Source: EEX (2003)

Figure 3-5. Electricity future prices

2.3   Availability of plants,

For thermal power plants, two major reasons for non-availability may be distinguished:
- Planned non-availability due mostly to maintenance and revision activities
- Forced outages following some incident during operation

Some illustrative numbers are given in Table 3-1. They clearly indicate that for thermal power plants availability is usually higher than 85 % and that the major part of non-availability is planned.

For hydro and other renewable power plants the non-availability of the natural energy supply constitutes a third, major cause of non-availability. Therefore, for these power plants an important distinction is between technical availability and overall availability (including resource availability). For renewable power plants with zero variable costs (hydro, wind, PV), the latter corresponds to the load factor and is strongly dependent on local conditions. In Germany, run-of-river hydro plants have on average a load factor[11] of 0.61 (cf. VDEW 2000), whereas for wind much lower availabilities are achievable, according to Ender (2003) corresponding to a load factor of 0.22.

Table 3-1. Typical availability factors for selected plant types

	Technical availability[a]	Planned unavailability	Non-planned unavailability
Coal-fired plants	85.0 %	10.0 %	5.0 %
Gas turbines	85.1 %	11.4 %	3.5 %
Nuclear plants	92.2 %	7.7 %	0.1 %

[a] so-called work availability accounting for partial availability in some periods

Source: VGB (1998)

2.4   Behavior of competitors

From the perspective of a single utility, the behavior of competitors has usually also to be considered as uncertain. Yet it is a different type of uncertainty than those discussed before. Since the behavior is at least most of the time driven by motivations of profit maximization, it is mostly uncertain in the view of others due to hidden (or proprietary) information or hidden (or proprietary) action. The case of hidden information and hidden

---

[11] In Germany, more frequently "full load hours" are used to characterize the ratio of annual output to installed capacity. These can be easily obtained by multiplying the load factor with 8760.

action will not be dealt with in depth here, but the market equilibria which may arise from the interaction of different profit maximizing actors will be at the heart of chapter 5.

2.5   Demand growth

Over the past years, electricity demand has grown only slowly in most OECD countries (cf. Figure 3-6). Yet this growth has been far from certain and subject to stochastic fluctuations as indicated in Figure 3-7. Especially in the longer run, uncertainty on electricity demand growth has therefore also to be taken into consideration.

2.6   Technology development

In the longer run, uncertainties on technology development are also of considerable relevance. An example are the future prospects of fuel cells or other distributed generation technologies One possibility for assessing the development trends is the use of learning or experience curves (e. g. Wene et al. 2000) to identify the likely development. But it is obvious that major uncertainties remain even when using learning curves – as illustrated in Figure 3-8 the learning rates observed in practice scatter around a wide range and as a consequence necessary cumulative production quantities for reaching a pre-specified target cost level may vary by more than a factor two. One might even argue philosophically (or more precisely epistemologically) that there must be a fundamental uncertainty when one human mind (the modeler) aims at assessing ex ante what will be the final outcome of the work of a multitude of other human minds (the fuel cell engineers). Ways of dealing with this type of uncertainty are discussed in chapter 8.

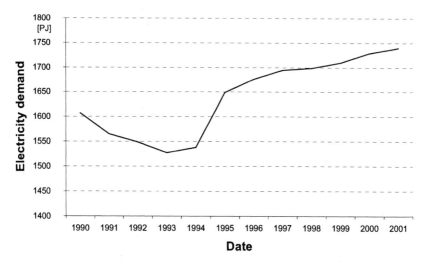

Source: AG Energiebilanzen (2002)

Figure 3-6. Development of electricity demand from 1990 to 2001 in Germany

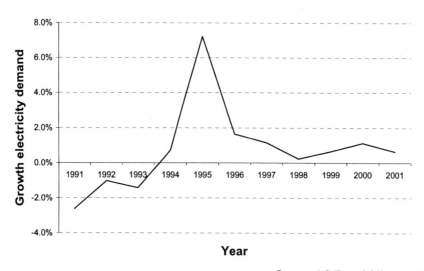

Source: AG Energiebilanzen (2002)

Figure 3-7. Growth rate for electricity demand from 1990 to 2001 in Germany

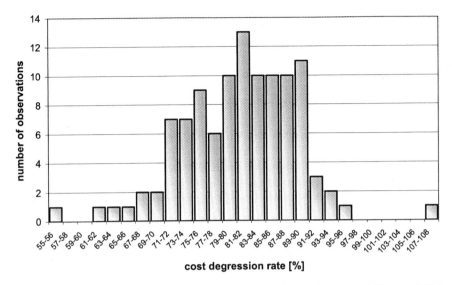

Source: Dutton and Thomas (1984)

Figure 3-8. Cost degression rates observed for various technologies

2.7   Regulatory and political uncertainties

Finally regulatory and political uncertainties are last type of uncertainty, which is of preeminent importance for the electricity industries. The last decade has shown that the changes in the regulatory framework are not finished once competition is introduced. On the contrary, the regulatory framework has continuously evolved in all competitive electricity markets in order to cope with perceived imperfections of the market design. Recent examples include the new trading arrangements (NETA) finally introduced in England and Wales after long discussions in 2001 (OFGEM 2002), the standard electricity market design (SMD) developed by the Federal Electricity Regulatory Commission (FERC 2002) in the U.S. after the California power crisis, or the different versions of the 'Verbändevereinbarung' (BDI et al. 1998, BDI et al. 1999, BDI et al. 2001) developed in Germany. To this, political interference has to be added which aims at promoting other policy goals, notably emission reduction and climate change mitigation through various instruments such as tradable emission certificates or renewables feed-in tariffs. This changeable political environment has had a huge impact on decisions in the electricity industry in the past and is expected to be also of considerable importance for future choices, e. g. through political decisions on the (non-)use of nuclear energy or on the promotion of renewables.

From the perspective of a single utility, the political uncertainties are in a certain sense more difficult to cope with than e.g. price uncertainties, since hardly any objective probabilities can be assigned to possible outcomes of the political process. In the subsequent chapters nevertheless some elements how to deal with such uncertainties will be discussed.

3. FORMAL FRAMEWORK

A first step to cope with the decision problems and the uncertainties described in the previous sections is to define some notation to handle such problems. This will be done in the following sections, starting with a categorization of decision support models in section 3.1, followed by the notation for describing uncertain parameters in section 3.2. Section 3.3 then uses this framework to formulate one basic decision problem in the electricity generation business: the unit commitment and load dispatch problem.

3.1 Formulating decision problems under uncertainty

In a static view[12], decision problems are essentially composed of the following elements:

- a set of decision alternatives $\mathbf{A} = \{A_k\}$
- a set of possible states of the world $\mathbf{S} = \{S_l\}$
- a results function $V(A_k, S_l)$
- a utility function or functional $J(V(A_k, S_l))$
- a decision criterion (objective function)

The following remarks have to be added to these basic definitions:

The set of decision alternatives $\mathbf{A}$ may have a finite number of elements (e.g. the two elements: to run or not a given plant) or an (almost) infinite number of elements (e.g. the possible bidding prices in a power auction).

The set of decision alternatives may be described by a vector $\mathbf{y}$ of decision variables. Without much loss of generality we can assume that $\mathbf{y}$

---

[12] A dynamic view on decision problems, which emphasizes the stages of problem identification, problem analysis, decision taking, implementation etc. is certainly also very useful for reaching good decisions in the practice of the electric industry. However, such a perspective on decision problems is largely outside the scope of this book, which focuses on providing methods for the stages of problem analysis and decision taking.

has a number $d_{y,int}$ of integer elements and a number $d_{y,re}$ of real-valued elements, i.e. $\mathbf{y} \in B^{d_{y,bin}} \times \Re^{d_{y,re}}$ with $B_n = \{1, ..., n\}$.

$\mathbf{y}$ may have a temporal structure, in this case a partition $\mathbf{y} = [\mathbf{y}_1 \, \mathbf{y}_2 \, ... \mathbf{y}_t \, ... \mathbf{y}_T]^T$ exists.

The set of possible states of the world may also have a finite or an infinite number of elements and may be described by a vector $\mathbf{x}$ of state variables. For $\mathbf{x}$ it is assumed that $\mathbf{x}$ has a number $d_{x,int}$ of integer elements and a number $d_{x,re}$ of real-valued elements, that is $\mathbf{x} \in B_n^{d_{x,bin}} \times \Re^{d_{x,re}}$ with $B_n = \{1, ..., n\}$. Furthermore, $\mathbf{x}$ may also have a temporal structure.

$\mathbf{x}$ can be described as a (vector of) random variable(s) (cf. below).

However, the description of $\mathbf{x}$ as random variable is not useful in all decision settings. In traditional decision theory, the distinction is made between decisions under uncertainty (in a narrow sense) and decisions under risk (cf. Bamberg, Coenenberg 1994). Decisions under risk cover those cases where objective probabilities can be assigned to the different possible states of the world (values of $\mathbf{x}$). In that case $\mathbf{x}$ can be described as random variables. Decisions under uncertainty (in a narrow sense) cover conversely the cases where no assignment of objective probabilities is possible[13].

Courtney et al. (1997) distinguish between four different types of uncertainty (in the narrow sense):

Level 1 – a clear enough future describes the case were despite some uncertainty decision makers can develop one single forecast as basis for decision development.

Level 2 – alternate futures corresponds to circumstances where the future can be described through a set of discrete scenarios. This can correspond to different policy regulations or to different strategies by competitors

Level 3 – a range of futures is characterized by the fact that there are no obvious discrete alternatives but rather a range of possible outcomes, e.g. in an emerging market the achievable customer-penetration may vary between 5 % and 40 %.

Level 4 – true ambiguity occurs if multiple dimensions of uncertainty interact to create an environment that is virtually impossible to predict. Such situations are according to Courtney et al. (1997) rather rare but they do exist.

---

[13] The subjective expected utility (SEU) theory first developed by Savage (1954) argues that subjective probabilities are sufficient to treat a decision problem using known probabilities. In the context of an electric utility, where most decisions are taken not by single individuals but by groups, the existence of consistent collective subjective probabilities can however not be taken for granted.

These different types of uncertainty will arise in the electricity industry especially in the longer term and we will therefore come back to appropriate methods for decision support when analyzing the longer term problems of technology development and investment strategies. In the next section we will turn towards decision settings where the unknown parameters may be described through probabilities.

3.2   Describing uncertain parameters

In mathematics, probability measures are introduced within the rather general framework of measure theory. A probability space is in that context a triple (Ω, Θ, Φ), where Ω describes the set of possible values. Θ is a σ-algebra of Ω, i.e. a set of possible subsets of Ω and Φ is a probability measure, i.e. a mapping from Ω on the interval $[0,1]$ with the following properties:

$$\Phi(\varnothing) = 0$$
$$\Phi(\Omega) = 1 \tag{3-1}$$
$$\Phi\left(\bigcup_{n=0}^{\infty} E_n\right) = \sum_{n=0}^{\infty} \Phi(E_n)$$

Thereby $(E_n | n \in N)$ is any sequence of pair-wise disjoint members of Θ. The last property describes the additivity of probabilities of disjoint events.

Based on these basic definitions, in a next step random variables and stochastic processes can be introduced.

3.2.1   Random variables

Random variables are defined as functions which map the set of possible events Ω to (a subset of) real numbers:

$$X : \Omega \mapsto \Re \tag{3-2}$$

X itself is not observable, but based on the properties of the probability space (Ω, Θ, Φ), properties of X can be deduced. If the image of X contains only a finite or countable infinite number of elements, discrete probabilities Pr_X can be assigned to these elements using the probability measure Φ defined on Θ.

$$Pr_X : \quad \begin{array}{l} \Re \mapsto [0,1] \\ x \rightarrow Pr_X(x) = \Phi(\{\omega | x = X(\omega)\}) \end{array} \tag{3-3}$$

For random variables X with a non-countable image set, the definition of discrete probabilities Pr_X is usually not possible. But always the cumulative distribution function (CDF) $F_X(x)$ can be defined through the probability measure Φ:

$$F_X : \quad \begin{array}{l} \Re \mapsto [0,1] \\ x \rightarrow F_X(x) = \Phi(\{\omega | X(\omega) \leq x\}) \end{array} \tag{3-4}$$

If F_X is (continuous and) differentiable, the probability density function (PDF) f_X may be defined as the derivative of F_X with respect to x:

$$f_X : \quad \begin{array}{l} \Re \mapsto [0,1] \\ x \rightarrow f_X(x) = \dfrac{df_X(x)}{dx} \end{array} \tag{3-5}$$

If f_x is defined, the expectation of the random variable X can be defined easily through:

$$E[X] = \int_{-\infty}^{+\infty} x f_X(x) dx \tag{3-6}$$

But as described for example in Breiman (1986), the expectation of random variables can also be defined in those cases where no probability density function exists for X. Since we will occasionally also use stochastic variables with non-continuous cumulative distribution functions, this definition is also given here:

$$E[X(\omega)] = \lim_{\delta \rightarrow 0} \lim_{m \rightarrow \infty} \left[\sum_{n=0}^{m} n \delta \Phi(\{\omega | n\delta \leq \omega \leq (n+1)\delta\}) \right] \tag{3-7}$$

As usual with real-valued random variables, it is thereby assumed that also the sample space Ω is $\Re$. If Ω is extended to be a Cartesian product of $\Re$, $\Re^T$ with $T \in N$ or even an infinite-dimensional space such as a sub-space of the real-valued functions, we get stochastic processes.

3.2.2   Stochastic processes

Stochastic processes **X** in discrete time can be most easily viewed as a set of random variables X_t indexed over time t.

$$\mathbf{X} = \left\{ X_t \middle| X_t : \Omega \mapsto \Re, t \in \{1,2,...,T\} \right\} \tag{3-8}$$

The possible values of X_t are often called states, the set of all possible values for X_t is called state space. Multivariate stochastic processes are sets of random variables indexed over time and further dimensions, like products, power plants, customers, etc. Multi-variate stochastic processes will be considered in more detail in chapter 6 on risk management.

In financial mathematics, stochastic processes in continuous time are also frequently used to describe price movements. Mathematical measure theory is particularly useful for deriving general properties of these processes. They can be viewed as a family of random variables:

$$X(t) = \left\{ X(t) : \Omega \mapsto \Re \middle| t \in [0\ T] \right\} \tag{3-9}$$

The use of stochastic processes in continuous time allows for selected problems to derive analytical solutions (e.g. the famous Black-Scholes formulas). But real-world decision problems in the electricity industry are in most cases more adequately formulated in discrete time and the models are often anyhow so complicated that no analytical solutions are possible. Therefore the properties of continuous time processes will only occasionally be used.

3.3   Models for decision support under uncertainty

In general, two types of models are of interest for the decision support in the electricity industry. On the one hand, adequate models are needed to describe the development of uncertain parameters, notably electricity prices or demand. These are *descriptive models*, since they focus on providing a description as good as possible of observed phenomena. For descriptive models, model quality has hence to be measured by comparing actual observations $\widetilde{X}_t$ with the stochastic variables X_t provided by the model specification. How this can be done is discussed in section 3.4.

On the other hand, decision support aims at providing answers to the question, which would be the best decision in a given context. This is the aim of *prescriptive* or normative *models*, although the term "prescriptive" has not to be taken literally, since the decision has always to be made by the decision maker and he or she may and should always consider the possibility that the decision proposed by the model is not optimal, given that some elements of reality are not fully reflected in the model. The performance of prescriptive models on real problems can only be measured by comparing its outcomes ex-post to the performance of some alternative model (which could be the "intuitive" decisions of the decision maker).

It has to be noted that the division between descriptive and normative models is in practice not as clear-cut as it may seem at first sight. Notably, if a descriptive model describes the behavior of optimizing agents or efficient markets, it might well be formulated as an optimization model, i.e. in the same formal structure than a prescriptive model (cf. chapter 4, section 1). Another somewhat different example is the value of an option, which is usually determined on the basis of a descriptive model, but which may be directly interpreted as the prescription not to pay more than the calculated value.

Yet as a general rule descriptive models are used to describe and predict the value of parameters which are outside of the decision maker's control but influence his or her decision (e.g. electricity price). Prescriptive models aim at identifying the optimal decision under these circumstances, making use of the information provided by the descriptive models.

3.4   Measuring model quality

In order to measure the quality of *descriptive models*, various statistics have been developed (cf. e.g. Hufendiek 2001). The most widely used are:

- The share of explained variance (goodness-of-fit measure)

$$R^2 = \frac{\sum_{t=1}^{T}\left(\tilde{X}_t - \hat{X}_t\right)^2}{\sum_{t=1}^{T}\left(\tilde{X}_t - \mu\left[\tilde{X}_t\right]\right)^2}$$

- The root mean squared error $RMSE = \dfrac{1}{T}\sqrt{\sum_{t=1}^{T}\left(\tilde{X}_t - \hat{X}_t\right)^2}$

- The mean absolute error $MAE = \dfrac{1}{T}\sum_{t=1}^{T}\left|\tilde{X}_t - \hat{X}_t\right|$

- The mean percentage error $MAPE = \dfrac{1}{T}\sum_{t=1}^{T}\left|\dfrac{\tilde{X}_t - \hat{X}_t}{\tilde{X}_t}\right|$

- The value of the likelihood–function $L = \prod_{t=1}^{T} f_{X_t}\left(\tilde{X}_t \middle| \tilde{X}_1 ... \tilde{X}_{t-1}\right)$

- The value of the log-likelihood–function

$$LL = \ln(L) = \sum_{t=1}^{T} \ln\left(f_{X_t}\left(\tilde{X}_t \middle| \tilde{X}_1 ... \tilde{X}_{t-1}\right)\right)$$

Hereby the first four measures uniquely focus on the predictive quality of the models by comparing the actual observations $\tilde{X}_t$ with the predictions $\hat{X}_t$ obtained from the model. Usually, the predictions are thereby set equal to the expected value of the stochastic variable: $\hat{X}_t = E[X_t]$. The use of the likelihood-function (or equivalently the log-likelihood-function) by contrast emphasizes the as good as possible replication of the actual distribution through the stochastic model[14].

All measures may be evaluated on the sample used for estimating the model parameters, or they may be calculated based on a comparison of out-of-sample forecasts performed by using the model(s)[15]. The latter is a more appropriate indication of the model quality, especially if it is done using similar forecast horizons than those to which the model is to be applied in practice. Yet its proper application requires estimating the model repeatedly and using it each time also for the corresponding forecasts.

Which of the aforementioned quality measures is most appropriate, depends of the use that is made of the forecast values. Ultimately the quality criterion used for the evaluation of the descriptive models should reflect the costs associated with incorrect forecasts. This is a strong argument to use measures of absolute error such as MAE or RMSE in the case of price forecasts instead of relative measures like MAPE – a forecast error of 5 % is much more important if the price is 100 €/MWh than if the price is 10 €/MWh. One might even go one step further and argue that if the value of the objective function is purely linear in the price (e.g. in the case of a simple profit function), the MAE is more appropriate than the RMSE, which accords more weight to strong outliers. But if the optimal decision does not

---

[14] Based on the log-likelihood function adjusted measures like the Akaike information criterion (AIC) or Schwarz' Bayesian information criterion (BIC) have been developed which allow comparing models with different numbers of variables (cf. Box, Jenkins, Reinders 1994, Sakamoto et al. 1986). A concise discussion why the AIC (or at equal model size the log-likelihood) is an adequate criterion of model quality is provided by Chi, Russel (1999). A general introduction to maximum likelihood estimation in econometrics is provided inter alia by Greene (2000).

[15] The likelihood is usually only calculated on the estimation sample, but the extension to the out-of-sample case is straightforward.

only depend on the expected (or most probable) value of the explained variable but on its whole distribution, then the quality of the descriptive model should be measured taking into account the full distribution, i.e. using the (log-)likelihood-function.

An important remark is that a comparison of models with different explained variables (e.g. prices and log of prices) cannot be done by applying the quality statistics directly to the models. Instead, the model outputs have to be transformed to the same explained variable in order to achieve a meaningful comparison.

For *prescriptive models*, quality measurement is easier on the one hand, since with the objective function a "natural" quality criterion exists. But as stated before, quality can here always only be measured in relation to some other decision model. And comparison is further complicated if these models are applied repeatedly and the later decisions depend on those taken previously. Then only the quality of whole decision chains can be compared against each other. For an appropriate comparison, it is then important to perform the comparison in different situations, such as low and high price periods. And obviously, the quality of the decisions not only depends on the quality of the prescriptive model (decision rule) but also on the quality of the descriptive models delivering information for the decision support.

Chapter 4

MODELLING ELECTRICITY PRICES

Electricity prices are as described in chapter 3 a key source of uncertainty and a key challenge for decision support in competitive electricity markets. In the past various approaches have therefore been developed to analyze and predict electricity market prices. They may be broadly divided into at least five classes (cf. Table 4-1). Among the theoretically founded models, fundamental models allow accounting well for the impact of power plant characteristics and capacities, for restrictions in transmission capacities and demand variations (cf. section 1). The financial mathematical models are more suited for coping with the volatility of electricity prices and are often used for option valuation and risk assessment purposes (cf. section 2). A third category of models is formed by game-theoretic approaches, which are particularly adequate for analyzing the impact of strategic behavior on electricity prices (cf. chapter 5). Besides these models with strong theoretical foundations, other more empirically motivated models are found: The fourth class of models, statistical and econometric time-series models, relate the fluctuations of electricity prices to the impact of external factors such as temperature, time of the day, luminosity etc. The stochastic aspect of electricity price formation is here acknowledged albeit often not dealt with in much detail. In fact this type of models is much complimentary to finance models, in that the statistical and econometric models deal in detail with possible explanatory variables for electricity price fluctuations whereas the finance models focus on the stochastic part of the price change. Finally also so-called "technical" analysis and expert systems can be mentioned as methods used especially by practitioners to anticipate price movements based on the analysis of past price developments. These model classes will not be discussed in detail here. Instead section 3 describes an integrated novel model combining the two aforementioned approaches of fundamental and financial models.

Table 4-1. Types of electricity price models

Model category	Examples
Theoretically founded models	
Fundamental models	Kreuzberg (1999), Starrmann (2000) and Müsgen, Kreuzberg (2001), Kramer (2002)
Finance models	Pilipovic (1998), Schwartz (1997), Johnson, Barz (1999), Barlow (2002)
Game theoretical models	Green, Newbery (1992), Bolle (1992), Bolle (2001), Jebjerg, Riechmann (2001), Weber (2001)
Empirically motivated models	
Statistical models	Erdmann, Federico (2001)
"Technical" analysis and expert systems	

1. FUNDAMENTAL MODELS

The basic idea of fundamental models is to explain electricity prices from the marginal generation costs. Examples of such models include Kreuzberg (1999), Starrmann (2000), Müsgen, Kreuzberg (2001), Kramer (2002), Kurihara et al (2002) and ILEX (2003). Many more fundamental models have been developed by consultants or the utilities themselves and are therefore not published. Often, fundamental models are also incorporated in more sophisticated game theoretic (Jebjerg, Riechmann 2000, Ellersdorfer et al. 2000, Hobbs et al. 2002) or stochastic models (Skantze et al. 2000, Barlow 2002). Especially in the latter papers, continuous approximations to the fundamental electricity price formation are however used.

In the following, first the basic principle of cost minimization is discussed, which leads to the derivation of the merit-order curve as simplest fundamental model. From this basic model, several extensions are possible to arrive at better approximations of the reality. These include notably the inclusion of import and export, hydro-storage and start-up costs as well as operation constraints (minimum up and minimum down times) and reserve power markets. How this can be done is discussed in the subsequent sections and then key challenges for obtaining valid fundamental market models are summarized.

1.1 Basic model: cost minimization under load constraint and merit order

If the electricity spot market operates efficiently, it should lead to an efficient system operation with minimal costs, satisfying all customer demands. If customer demand is taken as price inelastic (and there is much evidence of almost price-inelastic demand, cf. e.g. the figure for Alberta given in Barlow 2002), prices will equal to the marginal generation costs of the last unit needed to fulfill given demand. This well-known result can also be derived from a formal analysis. Here system operation is described by the following set of equations:

$$\min_{y_{u,t}} C_t = \sum_u C_{Op,u,t} \tag{4-1}$$

$$C_t = \sum_u C_{Op,u,t} = \sum_u p_{F,u,t} h_u y_{u,t} \tag{4-2}$$

$$\sum_u y_{u,t} \geq D_t \tag{4-3}$$

$$y_{u,t} \leq K_{PL,u} \tag{4-4}$$

Equation (4-2) describes the objective function as minimization of operation costs, which are determined by plant output $y_{u,t}$ multiplied by the fuel price $p_{F,u,t}$ and heat rate h_u. , Equation (4-3) provides the balance of supply and demand. It is formulated as an inequality since production will anyhow not exceed demand if production costs are strictly positive. The capacity constraint for each unit is finally given by Eq. (4-4). Together these equations form a classical linear programming problem, which may be written as:

$$\min_{y_t} c_t^T y_t \tag{4-5}$$
$$\therefore \quad A y_t \geq b_t$$

The solution of this problem will be always located on corners or edges of the hyperplane described by the inequalities in (4-5). Furthermore, according to the duality principle in linear programming, the solution is equivalent to the solution of the dual problem:

$$\max_{\psi_t} b_t^T \psi_t \tag{4-6}$$
$$\therefore \quad A^T \psi_t \leq c_t$$

A closer inspection of the dual problem reveals that the vector of shadow prices ψ_t contains as first element the shadow price of demand $\psi_{D,t}$ and as following elements the shadow prices of the capacities $\psi_{K,u,t}$. The dual problem may thus be written:

$$\max_{p_{D,t},p_{K,u,t}} \quad G_t = D \cdot \psi_{D,t} - \sum_u K_u \psi_{K,u,t} \qquad (4\text{-}7)$$

$$\psi_{D,t} - \psi_{K,u,t} \leq p_{F,u,t} h_u \quad \forall u \qquad (4\text{-}8)$$

Equation (4-7) implies that $\psi_{D,t}$ will be maximized and $\psi_{K,u,t}$ (the capacity rents of the units) minimized. From Eq. (4-8) it can be deduced that $\psi_{D,t}$ and $\psi_{K,u,t}$ will move in parallel as soon as the shadow price of demand exceeds the marginal generation costs of the unit u. Putting this into Eq. (4-7), it is evident, that the profit G_t will not be increased anymore as soon as the sum of capacities with positive shadow prices equals or exceeds demand D_t. Furthermore, Eq. (4-8) implies that the shadow price of demand will be (at maximum) equal to the variable costs of the cheapest unit where the shadow price of capacity $\psi_{K,u,t}$ is zero. Consequently the marginal system price, which corresponds to the shadow price of demand, will be equal to the marginal costs of the most expensive unit needed to satisfy the demand restriction. Graphically, this result of price formation according to the merit order curve is illustrated in Figure 4-1.

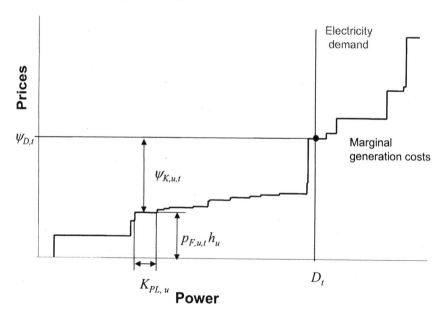

Figure 4-1. Merit order and shadow prices in the simple cost minimization model

Of course this is only a very basic model, which has to be extended into several directions, as shown in the following sections in order to cope at least approximately with the reality in European electricity markets.

1.2 Transmission constraints

A first extension to be considered is multi-regional modeling. If imports from region r' to r and exports from r to r' are allowed, the demand equation has to be formulated for each region and import and export flows e have to be added:

$$\sum_{u\in U_r} y_{u,t} + \sum_{r'}\left(e_{r',r,t} - e_{r,r',t}\right) \geq D_{r,t},\tag{4-9}$$

Transmission capacity restrictions are accounted for similarly to generation capacity restrictions:

$$e_{r,r',t} \leq K_{TR,r,r'} \cdot v_{TR,r,r',t}.\tag{4-10}$$

$$y_{u,t} \leq K_{PL,u} \cdot v_{PL,u,t},\tag{4-11}$$

Thereby, $D_{r,t}$ describes demand in region r, $e_{r',r,t}$ the export flows from region r' to r, $K_{PL,u}$ the capacity of power plant type u and $K_{TR,r,r'}$ the transmission capacity from region r to r'. Additionally, the availability of transmission lines $v_{TR,r,r',t}$ has been included into this extended model as well as the availability of power plants $v_{PL,u,t}$.

1.3 Hydro plants

Besides the thermal power plants focused on so far, hydro power plants also play a considerable role in many electric power systems including the continental European one. Here at least three cases have to be distinguished:
- Run-of-river plants
- Hydro storage plants
- Hydro storage plants with pumping facilities (pumped storage plants)

The first type of plant will run whenever water is available. It is therefore a must-run plant and its production $y_{u,t}$ can be set equal to the inflow $w_{u,t}$ or even directly be deduced from overall demand. Hydro storage plants are far more complicated, since through the storage, several time steps are linked together. This requires the formulation of the storage equations:

$$H_{u,t} \leq H_{u,t-1} - y_{u,t} + w_{u,t} \qquad \forall u \in \mathbf{U}_{STOR}, \qquad (4\text{-}12)$$

$$H_{u,t} \leq H_{MAX,u} \qquad \forall u \in \mathbf{U}_{STOR}, \qquad (4\text{-}13)$$

The first one describes the filling and discharging of the storage, the latter the maximum storage capacity constraint. The storage level $H_{u,t}$, which is expressed for simplicity in energy units, is furthermore as all other variables presumed to be positive.

The third class of hydro power plants, pumped storage plants, requires even further extension of the framework. Here, an additional decision variable, the pumping energy $\tilde{y}_{PUMP,u,t}$ has to be introduced, and Eq. (4-12) has to be modified accordingly:

$$H_{u,t} \leq H_{u,t-1} - y_{u,t} + \eta_{PS}\tilde{y}_{PUMP,u,t} + w_{u,t} \quad \forall u \in \mathbf{U}_{PS}, \qquad (4\text{-}14)$$

Thereby η_{PS} is the cycling efficiency of the pumped hydro plant, i.e. the fraction of the energy recovered when first pumping the water and then propelling it through the turbine again. Of course, the pumping energy has also to be included in the demand Eq. (4-9) and a capacity constraint on pumping has to be introduced:

$$\tilde{y}_{PUMP,u,t} \leq K_{PUMP,u} \cdot v_{PUMP,u,t}, \qquad (4\text{-}15)$$

In order to model adequately seasonal hydro storage, as practiced notably in the Alps, the model has to cover at least one year. Consequently the objective function (4-1) has to be extended to cover several time periods:

$$\min_{y_{u,t},\tilde{y}_{PUMP,u,t}} C_t = \sum_t \sum_u C_{Op,u,t} \qquad (4\text{-}16)$$

and all relevant restrictions for the operating variables have to be considered simultaneously. Furthermore, an adequate terminal condition has to be included for the water reservoirs. One attractive formulation is to require that the final and the initial reservoir level are identical, which can be expressed through the following cyclical condition for time step 1 (instead of Eqs. 4-12 resp. 4-14):

$$H_{u,1} \leq H_{u,T} - y_{u,1} + \eta_{PS}\tilde{y}_{PUMP,u,1} + w_{u,1} \quad \forall u \in \mathbf{U}_{PS}, \qquad (4\text{-}17)$$

1.4   Start-up costs and minimum operation/shut-down time

Start-up costs may influence considerably the unit commitment decisions of plant operators. In unit commitment and load dispatch models, they are typically modeled using binary variables for unit operation, start-up and shut-down (cf. chapter 6). However, this is hardly feasible when modeling a national or regional market. Nevertheless, an approximation can be done by defining as additional decision variable for each plant type the capacity currently online $\tilde{y}_{ONL,u,t}$. On the one hand, the capacity online then forms an upper bound to the output and on the other hand, the capacity online premultiplied by the minimum load factor r_{MLF} for the plant type gives a lower limit to the output:

$$y_{u,t} \leq \tilde{y}_{ONL,u,t}$$
$$r_{MLF}\,\tilde{y}_{ONL,u,t} \leq y_{u,t} \tag{4-18}$$

Start-up costs then arise, if the capacity online $\tilde{y}_{ONL,u,t}$ is increased, i.e. when the start-up capacity $\tilde{y}_{STU,u,t}$ gets strictly positive:

$$\tilde{y}_{STU,u,t} \geq \tilde{y}_{ONL,u,t} - \tilde{y}_{ONL,u,t-1} \tag{4-19}$$

In order to avoid that units are always kept online, one has to account for that the efficiency at part load is usually lower than at full load, i.e. instead of (4-2), the following equation describes the operation costs:

$$C_{Op,u,t} = p_{F,u,t}h_{0,u}r_{MLF}\tilde{y}_{ONL,u,t} + p_{F,u,t}h_{m,u}\left(y_{u,t} - r_{MLF}\tilde{y}_{ONL,u,t}\right)$$
$$= p_{F,u,t}h_{m,u}y_{u,t} + p_{F,u,t}\left(h_{0,u} - h_{m,u}\right)r_{MLF}\tilde{y}_{ONL,u,t} \tag{4-20}$$

Here $h_{0,u}$ is the heat rate at the minimum load factor and $h_{m,u}$ is the marginal heat rate between minimum and full load, which is assumed to be constant. If $h_{m,u} < h_{0,u}$, the operators have an incentive to reduce the capacity online.

Additionally, the cost function to be minimized has now to include also the start-up costs, described by the relation

$$C_{STU,u,t} = c_{STU,u,t}\tilde{y}_{STU,u,t} \tag{4-21}$$

For the modeling of minimum operation times, the variable $\tilde{y}_{ONL,u,t}$ may be used again. The requirement is that the reduction in the capacity online between time t and time $t+1$ cannot exceed the minimum of the capacity online during the last $T_{OP,MIN,u}$ hours. These hours correspond to the minimum operation hours of the corresponding plant type:

$$\tilde{y}_{ONL,u,t} - \tilde{y}_{ONL,u,t+1} \leq \tilde{y}_{ONL,u,\tau} \qquad \forall \tau \geq t - T_{OP,MIN,u} \qquad (4\text{-}22)$$

Conversely, the maximum start-up capacity is limited by the minimum of the capacity shut-down during the last $T_{SD,MIN,u}$ hours.

$$\tilde{y}_{STU,u,t} \leq K_{PL,u} - \tilde{y}_{ONL,u,\tau} \qquad \forall \tau \geq t - T_{SD,MIN,u} \qquad (4\text{-}23)$$

1.5   Reserve power markets

Depending on the national or international regulations, different types of reserve have to be provided by the generators. A rather detailed model of the different cases for the German market can be found in Kreuzberg (1999). A somewhat simpler model distinguishes only between spinning reserve (primary and secondary reserve) and non-spinning reserve (minute or tertiary reserve). In both cases, the positive reserve has to be included in the capacity balance of those plants able to provide them and additionally the negative spinning reserve margin has to be included as lower margin between the actual output and the minimum output:

$$\tilde{y}_{ONL,u,t} + \zeta_{RESsp+,u,t} + \zeta_{RESoth+,u,t} \leq K_{PL,u,t}$$
$$r_{MLF}\,\tilde{y}_{ONL,u,t} \leq y_{u,t} - \zeta_{RESsp-,u,t} \qquad\qquad (4\text{-}24)$$

An inclusion of a negative tertiary reserve is not required in systems with limited fluctuating generation (notably wind), since the tertiary reserve is usually required to cope with forced plant outages. In systems with high wind energy generation, a sudden reduction of conventional generation and hence a negative tertiary reserve may be required to cope with increased wind generation. This has to be accounted for in the constraint (4-22) for the minimum operation time, which has to be modified as follows:

$$\tilde{y}_{ONL,u,t} - \left(\tilde{y}_{ONL,u,t+1} + \zeta_{RESoth-,u,t}\right) \leq \tilde{y}_{ONL,u,\tau} \quad \forall \tau \geq t - T_{OP,MIN,u} \qquad (4\text{-}25)$$

Besides these restrictions at the plant level, which may include also maximum levels of reserve power to be provided by different types of plants, of course also the overall reserve restrictions have to be satisfied (cf. UCPTE 1998, DVG 2000):

$$\sum_{u \in \mathbf{U}_{sp+}} \varsigma_{RESsp+,u,t} \geq \varsigma_{RESsp+,ges,t}$$

$$\sum_{u \in \mathbf{U}_{sp-}} \varsigma_{RESsp-,u,t} \geq \varsigma_{RESsp-,ges,t}$$

$$\sum_{u \in \mathbf{U}_{oth+}} \varsigma_{RESoth,u,t} \geq \varsigma_{RESoth+,ges,t} \qquad (4\text{-}26)$$

$$\sum_{u \in \mathbf{U}_{oth-}} \varsigma_{RESsp+,u,t} \geq \varsigma_{RESsp+,ges,t}$$

1.6   Implications and challenges

The Eqs. (4-9) to (4-26) presented in the previous sections describe the price formation in an efficient and deterministic market for electricity. For the practical implementation of such a model, three major challenges arise. The first one is data availability. Depending on the market, more or less information on plant capacities and costs, demand patterns and transmission capacities may be available to construct such a model. A second challenge is the choice of appropriate time resolution. On the one hand, the modeling of seasonal hydro storage (cf. section 1.3) necessitates the modeling of a full year; on the other hand the adequate modeling of start up costs requires a time resolution of one hour or at most two. In order to keep the model manageable, one solution is to model typical days, as will be done within the framework of the integrated model (cf. section 3.2). Another is to use load segments within a seasonally decomposed yearly model (cf. Kreuzberg 1999). In this framework, the integration of start-up costs is however difficult.

A final challenge is the incorporation of stochastic fluctuations, e.g. in demand or plant availability. This is particularly relevant, if the model is to be used directly for short-term predictions (time horizon of up to two weeks). At longer time horizons, the impact of current values of the stochastic variables on the future prices is rather limited, as can be both explained fundamentally (stability of weather conditions, duration of plant outages) and observed empirically (cf. section 2). Of course, it is in principle possible to model all fundamental stochastic fluctuations also fundamentally, doing e.g. Monte-Carlo-simulations of demand variations, plant outages etc. But even if these processes could be modeled bottom-up perfectly, one should not expect to describe the full range of fluctuations observed on the markets: As demonstrated for the financial markets (cf. Hull 2000), trading itself is expected to contribute to the creation of stochastic fluctuations.

2.   FINANCE AND ECONOMETRIC MODELS

Describing the stochastics of price movements is the field par excellence of the numerous models developed in finance. Originally developed for stock and interest rate markets, quite a number of these models have also subsequently been applied to the energy field (cf. section 2.1). Econometric models are also dealt with in this paragraph, since there is considerable overlap between these two categories. The key emphasis of econometric models is on the inclusion and specification of deterministic regressors (cf. section 2.2). Yet given the specificities of electricity, notably its non-storability, not only specific models for the deterministic term have to be considered, but also the stochastics may be more adequately modeled by more advanced models. Relevant specifications hereof are discussed in section 2.3

2.1   Basic models

At the start, price movements on the wholesale markets for electricity and other energy carriers have been described using models originally developed for modeling the stock and interest rate markets. And even today the geometric Brownian motion is still used frequently to describe electricity market prices on the forward and future markets. Therefore it will be briefly reviewed in section 2.1.1, followed by the mean-reversion process in section 2.1.2. Even more relevant is the extension to jump-diffusion processes, discussed in section 2.1.3.

2.1.1   Geometric Brownian motion

Under geometric Brownian motion, rates of change (called return rates in the case of stocks and other equities) are described through generalized Wiener processes (cf. Hull 2000). A Wiener process is described in discrete time by the equation:

$$\Delta x_t = \varepsilon \sqrt{\Delta t} \qquad\qquad (4\text{-}27)$$

The change in the variable x during a small period Δt is hence the product of a random variable ε and the square root of the length of period

Δt. ε is so-called white noise, i.e. a random variable which has a standard normal distribution with an expected value of 0 and a variance of 1. A Wiener process is a specific form of a Markov-process, i.e. it is a stochastic process, where the current value contains all the information retrievable from the (price) history (cf. Hull 2000). The specification of stock markets as Markov processes is a direct consequence of the assumption of (weak) market efficiency, which implies that arbitrage trading will make current prices for a storable equity or commodity include all information available on future values. By looking at the limit $\Delta t \to 0$ and adding a (deterministic) drift term $\mu\, dt$, a generalized Wiener process is obtained:

$$dx = \mu dt + \sigma dz \tag{4-28}$$

Thereby μ is the drift rate of the generalized Wiener process and σ the volatility rate. Application to the price changes for electricity yields:

$$\frac{dp}{p} = \mu dt + \sigma dz \tag{4-29}$$

In discrete time, this may be written:

$$\frac{\Delta p_t}{p_t} = u = \mu \Delta t + \sigma \varepsilon \sqrt{\Delta t} \tag{4-30}$$

The relative price change u is hence the sum of a deterministic part $\mu \Delta t$ and of a stochastic part with variance $\sigma^2 \Delta t$.

The assumption of a geometric Brownian motion hence includes two important simplifications: firstly, the deterministic part of the price movement is presumed to depend solely on time, not on the price level or some exogenous factors. Secondly, the volatility of the stochastic part is assumed to be constant over time. Section 2.1.2 and section 2.2 will discuss ways to overcome the first limitation, whereas in section 2.1.3 and section 2.3 specifications going beyond the second limitation are presented.

Although the model of geometric Brownian motion is performing badly for energy and specifically electricity spot market prices (cf. Johnson, Barz 1999), it should be noted that the development of the corresponding future and forward prices may under certain circumstances be well described through a geometric Brownian motion, albeit possibly with time-varying volatility (cf. Clewlow, Strickland 2000, Karesen, Husby 2002).

2.1.2   Mean-Reversion models

Following the observations by Gibson and Schwartz (1990) and Johnson and Barz (1999) as well as others, it is usually assumed that energy market prices exhibit mean-reversion. The basic mean-reversion or Ornstein-Uhlenbeck process may be described in discrete time by the equation:

$$\Delta p_t = \kappa(p_0 - p_t)\Delta t + \sigma \varepsilon \sqrt{\Delta t} \qquad\qquad (4\text{-}31)$$

That is, the deterministic part of the price change Δp_t depends on whether the price p_t is currently above or below the equilibrium price p_0. With $\kappa > 0$ (otherwise the process is not stable), prices will increase on average if they are currently below the equilibrium price and vice versa. As before, the stochastic part is assumed to have expected value of zero and to be normally and independently identically distributed (i.i.d.).

The economic interpretation of the mean reversion observed for electricity prices is that many of the stochastic fluctuations having an impact on the electricity supply and demand balance are not permanent but only temporary, e. g. weather effects or outages. In this view, the equilibrium price p_0 represents an average price, which is modified through stochastic temporary effects. A sudden price shock will have no permanent effect in this model, since sooner or later (depending on the value of κ), the price will return to the equilibrium level. The opposite is true for a geometric Brownian motion: any price shock will have a permanent effect corresponding to its observed height, because all subsequent price changes are uncorrelated with the previous ones and hence have expectation zero.

Another important difference between geometric Brownian motion and mean reversion processes concerns the long-term price uncertainty. As shown in Figure 4-2, price uncertainty (measured as the standard deviation of the price change between t_0 and t) increases beyond any limit in the case of the geometric Brownian motion, whereas price uncertainty is bounded in the case of the mean reversion process[16]. This assumption might be rather problematic since it implies that price forecasts with a time horizon of 1 month - or any other value large compared to the time constant $1/\kappa$ - can be done with the same accuracy than price forecasts with a forecast horizon of 1 year.

---

[16] This difference is directly related to the distinction between instationary processes and stationary processes emphasized in econometric time series analysis (cf. e.g. Hamilton 1994).

Instead of formulating the mean-reversion model in the prices itself, it may also be formulated in the logarithms of prices (cf. Pilipovic 1998):

$$\Delta \ln p_t = \kappa \left(\ln p_0 - \ln p_t \right) \Delta t + \sigma \varepsilon \sqrt{\Delta t} \qquad (4\text{-}32)$$

This ensures that electricity prices are always positive and furthermore increases the probability of high prices. Nevertheless it may not be sufficient to describe the probability of price spikes observed in the electricity markets. This leads to the next model type, the jump-diffusion models.

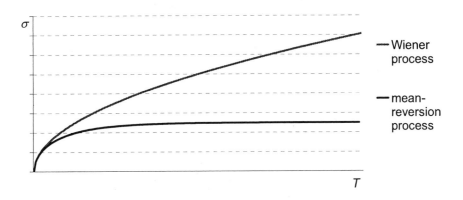

Figure 4-2. Price uncertainty as a function of forecasting horizon in Geometric Brownian motion and mean reversion processes

2.1.3 Jump-Diffusion models

Empirical investigations of the price changes in electricity markets clearly show that these are mostly not normally distributed, but that they rather have "fat tails", i.e. strong price jumps are more probable than under a normal distribution. This has been accounted for in finance and energy models through the addition of a jump process on the top of the conventional geometric Brownian motion or mean-reversion model. The original model proposed by Merton (1976) for stock prices is formulated in continuous time as:

$$dp = \mu p dt + \sigma_0 p dz + \sigma_s p \phi \, dq \qquad (4\text{-}33)$$

Thereby dq is a Poisson process, which gives as a counting process the probability of one or more jumps in a given time interval dt. The size of the jumps depends on the jump standard deviation σ_S and the normally distributed stochastic variable ϕ. Jump-diffusion models with mean reversion can as before be specified in the price itself or in its logarithm. Clewlow and Strickland (2000) propose the specification:

$$dp = \kappa(p_0 - p)dt + \sigma_0 p \, dz + \sigma_S p \, \phi \, dq \qquad (4\text{-}34)$$

For the estimation either a maximum-likelihood approach may be employed or the recursive filtering approach proposed by Clewlow, Strickland (1999). According to the authors, this is more suited to identify the extremely high spikes.

2.2   Deterministic and cyclical effects

Besides the specification of the stochastic structure, any spot market model for electricity has to account for the cyclical, more or less deterministic effects which are observable in spot market prices (cf. section 3.2.2). Three different types of cyclical effects can be distinguished: hour-of-the-day effects, day-of-the-week effects and seasonal effects. For all three, one of the following approaches could in principle be applied:

- Inclusion of continuous variables,
- Inclusion of corresponding dummy variables,
- Distinction of separate models.

For example Pilipovic (1998) uses sinus and cosinus functions to model seasonal effects, whereas Cuaresma et al. (2003) include monthly dummies to make the equilibrium price p_0 in the specifications (4-31) and (4-32) time dependent. Lucia and Schwartz (2000) test both specifications. Although monthly dummy variables seem at first sight more accurate, they are problematic when it comes to forecasting beyond the limits of one month: a sudden change in the time varying mean (estimated from the observed data) may induce a sudden price shock which is amplified through the mean-reversion process.

For the hour-by-hour variations, Cuaresma et al. (2003) test both a specification with dummy variables and a distinction of separate models as proposed by Ramanathan et al. (1997). They find clear evidence that a distinction of different models leads to a higher model quality, as measured through RMSE and MAE statistics. This is not surprising, since the inclusion of dummy variables allows only for time-varying means, yet separate models additionally lead to time-dependent volatilities and/or time-dependent mean-reversion rates.

Yet separate models are not always feasible. Notably distinguishing models for winter and summer (or even for quarters) would considerably restrict the database for each single model and create severe continuity problems at the connection dates. But in addition to the evidence provided by Cuaresma et al. (2003), a distinction of models for weekdays and weekends should be envisaged, since both price and load patterns differ substantially between these day categories.

Besides these pure time effects, one may also wish to include explicitly the impact of some exogenous factors with known impact on electricity prices such as meteorological conditions like outside temperature and wind speed (cf. Erdmann, Federico 2001) or system characteristics like number of stations online (cf. Elliott et al. 2000). This will improve the forecasting abilities of the model if the exogenous factors can in turn be predicted with some accuracy. For both factors mentioned, forecasting is only possible in the range of a few days or at maximum some weeks. Beyond this limit, the exogenous factors can only be replaced for forecasts by their respective long-term (possibly time-varying) mean. Consequently, the overall model quality in terms of long-term predicting ability will hardly be improved by including them.

2.3   Advanced stochastic models

Besides the most common stochastic specifications discussed in section 2.1, also various other specifications have been proposed to cope with electricity price volatility. In the following, the particular strengths and weaknesses of ARMA and GARCH models as well as Mixture distributions, Markov regime switching, transformed diffusion models and multifactor models will be briefly reviewed.

2.3.1   ARMA models

The mean-reverting process discussed in section 2.1.2 is in fact in a discrete time framework equivalent to an autoregressive process of order one (AR(1) process). It can thus be interpreted as one (simple) member of the general class of ARMA models, which have been popularized through the work of Box and Jenkins (cf. Box et al. 1994). General ARMA models may be specified as:

$$x_t = \alpha_0 + \sum_{k=1}^{m} \alpha_k x_{t-k} + \sum_{l=1}^{n} \beta_l \varepsilon_{t-l} + \varepsilon_t \qquad (4\text{-}35)$$

They allow accounting for rather general dynamics and impacts of past observations on today's prices. The parameters α_k describe the autoregressive influence on current prices, whereas the parameters β_l express the impact of past errors on today's prices. One should note that in general an ARMA process implies, that the time series does not fulfill the martingale property, i.e. that the expected value for tomorrow equals the price of today. Furthermore, general ARMA processes have no Markov property, so not all information on future prices is contained in the current price value. For the estimation of the parameters of a general ARMA process, maximum likelihood has to be used – in the case of pure autoregressive (AR) processes, the least square method will provide consistent but not efficient estimates (cf. Greene 2000).

2.3.2   GARCH models

The experience in energy and other commodity and equity markets shows that phases with small price changes alternate with phases with higher price changes. In the latter phases, the markets are usually nervous and strong price jumps occur several days in a row. This implies that the conditional probability for a large price change is higher, if there has been such a price movement in the previous days. Such a price behavior may be described mathematically through so-called GARCH models (Generalized Autoregressive Conditional Heteroscedasticity) (cf. Bollerslev 1986, Bollerslev 1990). In this case, the variance σ_t^2 of the relative price change u_t of a product is described by:

$$\sigma_t^2 = \omega + \sum_{p=1}^{P} \alpha_p \sigma_{t-p}^2 + \sum_{q=1}^{Q} \beta_q u_{t-q}^2 \qquad (4\text{-}36)$$

Hence, the variance at time t is composed of a constant term ω, the autoregressive influences $\alpha_p \sigma_{t-p}^2$ of the variances of the previous time steps $t\text{-}p$ and the impact of price changes at previous time steps $t\text{-}q$ $\beta_q u_{t-q}^2$.

In practice, GARCH (1,1) models are most widely used[17]. Then Eq. (4-36) reduces to:

$$\sigma_t^2 = \omega + \alpha \sigma_{t-1}^2 + \beta u_{t-1}^2 \qquad (4\text{-}37)$$

---

[17] There exist also several non-linear extensions of GARCH such as Nelson's (1991) EGARCH, which shall not be discussed here in detail (cf. Campbell (1997), Söderlind 2003). An empirical application to energy markets of some of them can be found in Cuaresma et al. (2003).

The parameters α, β and ω describe the relative weights of the different influences. All parameters have to be positive in order to ensure that σ_t^2 is positive. Also the inequality $\alpha + \beta < 1$ must hold for a stationary model. The exponential moving average model, proposed by J.P. Morgan (2001) for VAR calculations, can be viewed as a particular GARCH model, where the following parameter values have been selected: $\omega = 0$, $\alpha = 0.95$ and $\beta = 0.05$.

For the empirical parameter estimation, Eq. (4-37) is preferably transformed to:

$$\sigma_t^2 = (1 - \alpha - \beta)\sigma_m^2 + \alpha\sigma_{t-1}^2 + \beta u_{t-1}^2 \qquad (4\text{-}38)$$

Thereby, σ_m^2 is the average variance of the observed time series (cf. Hull 2000). By this transformation, the number of parameters for estimation is reduced to two per time series. The parameter estimation has to be done again through the maximum-likelihood method.

The corresponding likelihood function $L(\alpha,\beta)$, which describes the probability for the observation of the historical time series, is then:

$$L(\alpha,\beta) = \prod_{i=1}^{T}\left[\frac{1}{\sqrt{2\pi\sigma_t(\alpha,\beta)^2}}\exp\left(\frac{-u_i^2}{2\sigma_t(\alpha,\beta)^2}\right)\right] \qquad (4\text{-}39)$$

Thereby σ_t^2 is calculated using Eq. (4-37), resp. (4-36) in the more general case. The corresponding log-likelihood function $LL(\alpha,\beta)$ (cf. section 3.3.4) is then:

$$LL(\alpha,\beta) = \frac{1}{2}\sum_{i=1}^{T}\left[-\ln(2\pi) - \ln\sigma_t^2 - \frac{u_i^2}{\sigma_t^2}\right] \qquad (4\text{-}40)$$

The maximization of the log-likelihood function has to be done numerically, since no analytical solution to the first order conditions exists. Here the conventional Newton-Raphson algorithms for numerical maximization of continuous functions may be applied or the problem-specific Levenbeg-Marquardt-algorithm (cf. Press et al. 1988, Hull 2000).

2.3.3 Mixture distributions

The unconditional distribution of price changes obtained through GARCH models exhibits „fat tails", yet the conditional distribution for the next day still is normally distributed. This is in contradiction with many observations (cf. e.g. Weber, Hoeck 2003). Besides modeling jump-diffusion processes, another way to overcome this shortcoming is to approximate the real distribution through a mixture of normal distributions. It can be shown that any distribution may be approximated to any given degree of accuracy through a sequence of normal distributions (cf. Ferguson 1973, Ferguson 1983, Geweke, Amisano 2001). A mixture of n normal distributions is thereby described in discrete time through:

$$u_t = \sum_{k=1}^{n} 1_{S_t=k} \, X_{k,t} \qquad X_{k,t} \sim N(\mu_k, \sigma_k); S_t \sim I(1:n, \text{Pr}) \qquad (4\text{-}41)$$

S_t is thereby an integer random variable with fixed probabilities Pr_k for each value k and describes the occurrence of the n states. The random variables $X_{k,t}$ are all normally distributed but with different means and standard deviations. Economically every state or regime k should be interpreted as a different type of market price behavior.

If the number n of regimes equals two, the mixture model can be shown to be equivalent to the jump diffusion model of section 2.1.3, if the maximum number of jumps per time steps is set equal to 1. In general, the probability density function of u_t may be written:

$$f_{u_t}(x) = \sum_{k=1}^{n} \text{Pr}_k \, f_k(x) \qquad (4\text{-}42)$$

An interesting application, which will be discussed further in chapter 7, is to distinguish three regimes:
- *Regime 0:* no price change
- *Regime 1:* ordinary price changes (normal market)
- *Regime 2:* sudden price jumps (nervous market)

This novel mixture model is particularly attractive for forward or future markets with infrequent trading, such as observed on many continental European market places. Every price regime k is itself normally distributed with different standard deviations (and possibly different means) and occurs with a certain probability Pr_k. The probability density function (4-42) becomes:

$$f_{u_t}(x) = \Pr_0 \delta_0(x) + \Pr_1 f_1(x) + (1 - \Pr_0 - \Pr_1)f_2(x) \qquad (4\text{-}43)$$

Thereby, $f_1(x)$ and $f_2(y)$ describe the normally distributed probabilities of the price changes in the ordinary (regime 1) and the nervous market (regime 2). For regime 0, a normal distribution is not exactly obtained, since the case "no price change" corresponds to a standard deviation of zero. Here the normal distribution shifts from its ordinary bell shape into an infinitely high, but infinitely small peak. This can be described by the so-called Dirac impulse δ, widely used in control theory (e.g. Föllinger 1985, p. 22). It is defined as follows:

$$\delta(x) = \begin{cases} +\infty & if \quad x = 0 \\ 0 & if \quad x \neq 0 \end{cases} \qquad \int_{-\infty}^{+\infty} \delta(x)dx = 1 \qquad (4\text{-}44)$$

For the parameter estimation, again the maximum-likelihood (ML) method is used. The corresponding log-likelihood function is:

$$LL(\Pr_0, \Pr_1, \mu_1, \sigma_1, \mu_2, \sigma_2) = \sum_t \ln f_{u_t}(x_t)$$

$$= \sum_t \ln(\Pr_0 \delta(x_t) + \Pr_1 f_1(x_t; \mu_1, \sigma_1) + (1 - \Pr_0 - \Pr_1)f_2(x_t; \mu_2, \sigma_2)) \qquad (4\text{-}45)$$

A difficulty for the estimation arises from the fact that the Dirac impulse obviously has a singularity for $x = 0$. Snoussi and Mohammad-Djafari (2001) have proposed an indirect log-likelihood procedure to circumvent this issue. A more direct approach is to use a discrete approximation to the Dirac impulse, which is not singular and which has the same observable properties as the true Dirac impulse. Since in any market there is a lower limit Δp_{min} for price differences (e.g. 0.01 €ct), the Dirac impulse may be replaced by a rectangle of width Δp_{min} and height $1/\Delta p_{min}$ for estimation purposes. Thus the log-likelihood function becomes:

$$LL(\Pr_0, \Pr_1, \mu_1, \sigma_1, \mu_2, \sigma_2) = \sum_t \ln f_{u_t}(x_t)$$

$$= \sum_t \ln\left(\Pr_0 \frac{1}{\Delta p_{min}} 1_{x=0} + \Pr_1 f_1(x_t; \mu_1, \sigma_1) + (1 - \Pr_0 - \Pr_1)f_2(x_t; \mu_2, \sigma_2)\right)$$

$$(4\text{-}46)$$

Again the value of the log-likelihood function may be computed as a function of the parameter values, and the best-fitting parameter values can once more be determined through numerical maximization. As a result, beside the probabilities P_0 and P_1 also the means μ_1, μ_2 and the standard deviations σ_1, σ_2 of the price regime are obtained.

2.3.4   Markov regime switching

The major drawback of the mixture distribution models is that - contrarily e.g. to the GARCH models - no dynamic impact of the volatilities of previous days are included in the model. The probabilities of the different regimes are fix and consequently independent of the price changes observed in the previous period. Hence this model is unable to replicate persistent phases of high or low volatility. Yet this limitation can be overcome through the so-called *switching regime approach* (cf. e.g. Hamilton 1989, Kim, Nelson 1999). It is based on a mixture model, but now the probabilities of occurrence of the different regime are no longer fixed, but depend on the previously observed price patterns. Consequently, the probability density function is no longer unconditionally valid according to Eq. (4-42). Instead the probability density function is now "conditional" or dependent on the regime, which is described through a status variable s_t:

$$f_{u_t}(x_t|s_t) = \sum_{k=1}^{n} 1_{s_t=k} f_k(x_t) \qquad f_k(x_t) \sim N(\mu_k, \sigma_k) \qquad (4\text{-}47)$$

In the specific case of three regimes discussed before, one gets[18]:

$$f(x_t|s_t) = \begin{cases} \delta_0(x_t) & if & s_t = 0 \\ f_1(x_t) & if & s_t = 1 \\ f_2(x_t) & if & s_t = 2 \end{cases} \qquad (4\text{-}48)$$

Thereby the probabilities for the regimes s_t at time t are not fix as under the mixture model, but dependent on the regime s_{t-1} from the previous time step. This Markov switching is assumed to occur with constant transition probabilities and under the conjecture of (weak) market efficiency, i.e. all information relevant for market prices at time t is contained in prices and

---

[18] Applications of Markov Switching models to Spot market prices can be found in Elliott et al. (2000), Huisman, Mahieu (2001), de Jong, Huisman (2002).

states at t-1[19]. In order to describe the link between s_{t-1} and s_t, the transition matrix $\mathbf{T}$ of size $n \times n$ is defined:

$$\mathbf{T} = (T_{kl}) \qquad with: T_{kl} = \Pr(s_t = k|s_{t-1} = l) \qquad (4\text{-}49)$$

Each element T_{ij} describes hence the probability that the regime $s_{t-1} = j$ is followed by regime $s_t = i$. Given the restriction that the sum of transition probabilities for one given state $s_{t-1} = j$ has to be equal to 1, the matrix $\mathbf{T}$ contains $(n\text{-}1) \times n$ independent elements. For the example discussed before, one gets:

$$\mathbf{T} = {}^{s_t}\left\{\begin{bmatrix} \Pr_{00} & \Pr_{01} & \Pr_{02} \\ \Pr_{10} & \Pr_{11} & 1-\Pr_{02}-\Pr_{22} \\ 1-\Pr_{00}-\Pr_{10} & 1-\Pr_{02}-\Pr_{22} & \Pr_{22} \end{bmatrix}\right. \qquad (4\text{-}50)$$

$$\overbrace{\qquad\qquad\qquad}^{s_{t-1}}$$

So the columns correspond to the regimes [0 1 2] of yesterday, whereas the rows describe the regimes [0; 1; 2] of today. Consequently the diagonal contains the probabilities that a prevailing price regime remains stable from one day to the next. E.g. $\Pr_{11}$ is the probability that a normal price change is followed by another normal price change.

Recursively, the probability of a regime i at time t can be deduced from:

$$\Pr(s_t = k|s_{t-1}) = \sum_{l=1}^{n} T_{kl} \Pr(s_{t-1} = l) \qquad (4\text{-}51)$$

This is the so-called Hamilton filter (Hamilton 1994b). In the example discussed so far, the probability for regime 1 at time t is consequently equal to:

$$\Pr(s_t = 1|s_{t-1}) = T_{10} \Pr(s_{t-1} = 0) + T_{11} \Pr(s_{t-1} = 1) + T_{12} \Pr(s_{t-1} = 2) \qquad (4\text{-}52)$$

One has thereby to be aware that in most cases no unequivocal statement about the current regime is possible, since the status variable s_t is "non

---

[19] Note that sometimes Markov switching is defined to include also processes with time lag larger than 1 (e.g. Kim, Nelson 1999), so that in that nomenclature the model class analyzed here has to be labeled Markov switching models with time lag 1.

observable" (at least not exactly). Even in a regime with high price volatility, small price changes will occur from time to time or even no price change might happen, just with a lower probability.

The likelihood-function is in this case the product of the a-priori probability density functions for each observation (Kim, Nelson 1999). It can hence be written:

$$L(\mathbf{T}, \boldsymbol{\mu}, \boldsymbol{\sigma}) = \prod_t \left(\sum_k f_k(x_t | s_t = k) \Pr(s_t = k | s_{t-1}) \right) \qquad (4\text{-}53)$$

Correspondingly the log-likelihood is:

$$LL(\mathbf{T}, \boldsymbol{\mu}, \boldsymbol{\sigma}) = \sum_t \ln \left(\sum_k f_k(x_t | s_t = k) \Pr(s_t = k | s_{t-1}) \right) \qquad (4\text{-}54)$$

For the example considered, one gets:

$$
\begin{aligned}
LL&\left(T_{00}, T_{10}, ..., T_{22}, \mu_1, \sigma_1, \mu_2, \sigma_2\right) \\
&= \sum_t \ln\left(\delta_0(x_t)\Pr\left(s_t = 0|s_{t-1}\right) + f_1(x_t; \mu_1, \sigma_1)\Pr\left(s_t = 1|s_{t-1}\right)\right. \\
&\qquad\qquad\qquad\qquad\left. + f_2(x_t; \mu_2, \sigma_2)\Pr\left(s_t = 2|s_{t-1}\right)\right) \\
&\hspace{10cm} (4\text{-}55)
\end{aligned}
$$

The distribution of the price changes is now a conditional distribution, depending on the price regimes in the past, and the probabilities of the price regimes are depending, according to Eq. (4-51), from the probabilities in the previous period. Consequently the arrival probabilities of the regimes may vary from one point in time to the next, even though the model takes as starting point constant transition probabilities between the regimes.

As before, parameter estimation can be done through numerical optimization of the log-likelihood-function (4-54). The parameters to be determined include now besides the regime means and standard deviations also the transition probabilities between regimes.

This requires the calculation, at a given transition matrix T, of the regime arrival probabilities $\Pr(s_t = k | s_{t-1})$ for each day, based on the probabilities of the preceding day, using Eq. (4-51). From these probabilities, the terms of the log-likelihood function are computed. By inserting the actual price changes therein, the a posteriori probabilities

$\Pr\left(s_t = k \mid s_{t-1}, x_t\right)$ are determined. These are in turn used in the following period for the determination of the a-priori probabilities for s_{t+1}. Hence a recursive computation procedure for each evaluation of the log-likelihood function is established. The optimization algorithm overwhelms this recursive procedure and finally derives those parameter values, which lead to an overall maximum for the log-likelihood according to (4-54).

2.3.5   Transformed diffusion model

Barlow (2002) uses the inverse ϑ_α of the Box-Cox transformation (cf. Box, Cox 1964), to specify a mean-reversion model, which contains the linear and the log-linear specifications of Eqs. (4-31) and (4-32) as special cases. He writes:

$$dX_t = \kappa\left(X_0 - X_t\right)dt + \sigma dz$$
$$p_t = \begin{cases} \vartheta_\alpha(X_t) \ if & 1 + \alpha X_t > \varepsilon_0 \\ \varepsilon_0^{1/\alpha} & if & 1 + \alpha X_t \leq \varepsilon_0 \end{cases} \tag{4-56}$$

with the power transform (inverse Box-Cox transformation):

$$\vartheta_\alpha(x) = (1 + \alpha x)^{1/\alpha} \ if \quad \alpha \neq 0 \ ; \ \vartheta_\alpha(x) = e^x \quad if \ \alpha = 0 \tag{4-57}$$

For $\alpha = 1$, the conventional Ornstein-Uhlenbeck (or mean-reversion) process is obtained, whereas $\alpha = 0$ corresponds to the log-linear model of Eq. (4-32). If $\alpha < 0$, the function ϑ_α increases more rapidly than an exponential function and the model produces more spikes than the log-linear specification. The model can again be estimated through non-linear maximum-likelihood methods, where some care has to be given to the numerical computation of the inverse ϑ_α. In his empirical application Barlow (2002) finds in almost all cases that the parameter α is significantly below zero, indicating a strong case for this "non-linear Ornstein-Uhlenbeck process" as compared to conventional linear or log-linear mean-reversion specifications.

An important advantage of this approach compared to alternative methods for tackling the spikes in the spot price series is the limited number of parameters, which has to be estimated. An inconvenience is however that the power transform (4-57) is only a crude approximation to the real merit order curve as depicted e.g. in Figure 4-1. Another drawback, shared by this model with many others presented before, is that it yields stationary time-series. I.e. like for the conventional mean-reversion model, the price uncertainty does not increase substantially with longer forecast horizons.

2.3.6   Multifactor models

When analyzing the price developments for storable commodities like oil or copper, multi-factor models where essentially introduced to cope with the empirically observed patterns of future price volatility (cf. Gibson and Schwartz 1990). Indeed neither the assumption of a geometric Brownian motion nor of a mean-reversion model for spot prices leads to satisfactory price patterns for the corresponding commodity futures. By applying a non-arbitrage argument, one can show (cf. e.g. Hull 2000, pp. 74f) that the future price p_F today must be equal to the expected spot price $p(T)$ at the expiration time T of the future, discounted with an appropriate interest rate i:

$$p_F(p,T) = e^{-iT} E[p(T)] \qquad (4\text{-}58)$$

Since in the case of Brownian motion, all price shocks have permanent effects, the volatility of the future price must be equal to the volatility of the spot price corrected by the factor e^{-it}, which is close to unity.

The opposite holds for the mean-reversion process. Given that prices are reverting to the mean, the long-term expectation for the spot price is equal to the mean price. This leads, as shown e.g. by Karesen and Husby (2002), to an exponentially decaying pattern for future prices as function of time to delivery, thus the volatility of long-term future prices should be close to zero.

Empirically, none of both is observed and this directs towards the hypothesis that future price volatility is not only influenced by spot market price volatility but also by one or several other factors. In the case of storable commodities, the second stochastic factor is usually interpreted, following the work of Gibson and Schwartz (1990) as convenience yield, i.e. loosely the value (or cost) of detaining physically the commodity under question instead of owning just the financial equivalent (on that concept see also Fama, French 1987). For an essentially non-storable commodity like electricity, the rationale for the concept of convenience yield is questionable, since at the moment of delivery there is a fundamental difference between owning the physical good and owning a contract on it. Therefore, the interpretation used by Pilipovic (1998) in her two factor model of the second factor being a longer-term, unobservable equilibrium price, seems more plausible. Over the past years, a multitude of two- and multi-factor models has been developed. One example is the two-factor model by Schwartz (1997):

$$dp = (\mu - \delta)pdt + \sigma_1 pdz_1$$
$$d\delta = \kappa(\delta_0 - \delta)dt + \sigma_2 dz_2 \tag{4-59}$$

where δ is the stochastic convenience yield, which follows a mean-reverting process and is interpreted as a (continuous) dividend on the original price. The processes dz_1 and dz_2 are assumed to be stochastic white noise, with correlation coefficient ρ. Another example is the model by Pilipovic (1998):

$$dp = \alpha(w - p)dt + \sigma_1 pdz_1$$
$$dw = \mu w dt + \sigma_2 w dz_2 \tag{4-60}$$

Here the white noise processes dz_1 and dz_2 are assumed to be uncorrelated.

Besides the question of interpretation of the additional stochastic factors, the multi-factor models also raise the question of identification of these parameters. Gibson and Schwartz (1990) as well as Karesen and Husby (2002) explicitly point at the need of using Kalman filtering techniques for identifying correctly the parameters of the models they propose. This is necessary since one of the variables entering the price equation *(δ resp. w)* is not directly observable, but also subject to perturbations (measurement errors in the terminology of state space models, cf. Harvey 1989). The models hence include a so-called latent variable which may only be identified approximately. If the disturbances in the measurement equation (the lower equation in 4-59 resp. 4-60) are rather high, it might even be impossible to determine adequately the latent variable. Gibson and Schwartz (1990) explicitly propose to use the nearest future contract instead of the spot market price, given that the latter is difficult to observe in the case of oil. Yet if this method is used for electricity prices, much of the short-term fluctuations are eliminated a priori. Karesen and Husby (2002) explicitly focus on the modeling of electricity spot prices and they use Nordpool data for their empirical application – but due to the large share of hydro generation, Nordpool is one of the electricity markets with the lowest intraday fluctuations (cf. section 3.2.2). Therefore it remains questionable, whether the multi-factor models proposed so far can be successfully applied to electricity spot market prices[20], especially given that spot price changes in most markets are not normally distributed, whereas the multi-factor models

---

[20] Also Clewlow and Strickland (2000, p. 123) fit their multi-factor spot price model to forward price data.

throughout assume normally distributed disturbances. Yet the basic idea behind multi-factor models, that electricity spot price movements in the longer run should be decomposed in stationary and non-stationary parts, seems worth retaining and one key feature of the integrated modeling approach proposed in the next section is precisely that it combines stationary and non-stationary price processes.

3.   INTEGRATED MODELING APPROACH

The previous discussion of fundamental and financial models for electricity price modeling has pointed at the relative strengths and weaknesses of both approaches. In particular the case for pure financial models is weakened by the fact that the non-arbitrage argument underlying many financial model developments can hardly be applied to electricity prices. Instead a time series of electricity spot prices has to be viewed as a series of consecutive but only partly linked market equilibria. These market equilibria may be computed (at least approximately) through fundamental models – but these have huge difficulties in describing the stochastics observed in the electricity market. Therefore in the following an integrated model is proposed, which combines fundamental and finance type models. More precisely, it uses the price established by a fundamental model as equilibrium price for a mean-reversion stochastic model.

The overall approach for this integrated model of electricity markets is sketched in Figure 4-3. In a first step, the stochastic development of prices on the primary energy market is modeled through a finance-type model (cf. section 3.1). The resulting energy prices are taken as an input to a fundamental model of the European electricity market described in section 3.2. This model yields marginal generation costs differentiated by time of the day, type of day and month in the year. These prices could be used as an input to a game-theoretic model yielding the prices and mark-up charged by strategic players in the market. However, such a model is not easy to solve in the proposed context (cf. chapter 5). Therefore, the system marginal costs are directly used as an input for a stochastic model of the electricity market (cf. section 3.3).

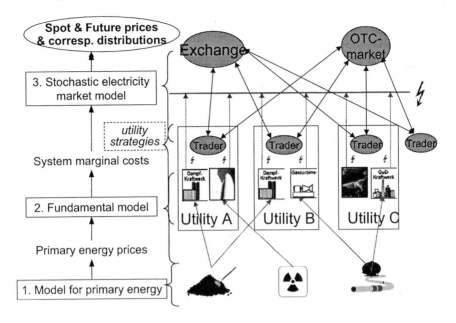

Figure 4-3. General approach for the integrated electricity market model

3.1 Primary energy prices: stochastic model

Many of the financial market models discussed in section 2 can also be applied for modeling the prices of primary energy carriers for electricity production. However, an important characteristic of the primary energy markets for oil and gas is that periods of high price volatility are alternating with periods of rather stable prices. This can be dealt with using a model with switching regimes, such as the one proposed by Erdmann (1995). Yet the moments of regime switching can be hardly determined a priori and therefore it is proposed as an alternative to model the primary energy market prices through a mean-reversion process for the derivatives of logs of prices $x_f = \ln p_f$:

$$d(dx_f) = \kappa_f \left(\mu_{df} - dx_f\right) + \sigma_f dz_f \qquad (4\text{-}61)$$

For the long-term properties of fuel price changes and corresponding electricity price shifts, the correlations between price changes for various energy carriers are furthermore important. As shown in Table 4-2, the long-term correlations between the different fuel prices are very high and significant. The short-term correlations on the contrary are considerably lower and partly far from being significant. Thus a straight-forward extension of model (4-61) to the multivariate case including just some correlations between the error terms will lead to a model which does not reproduce the long-term correlation pattern shown in Table 4-2.

Table 4-2. Correlation of primary energy prices and price changes

	Long-term correlations – correlations of logs of price					*Short-term correlations – correlations of changes in logs of prices*			
	coal	crude oil	natural gas	nuclear fuels		coal	crude oil	natural gas	nuclear fuels
coal	1	0,898*	0,814*	0,844*	Coal	1	0.344*	0.202	0.265
crude oil	0,898*	1	0,736*	0,892*	crude oil	0.344*	1	0.011	0.217
natural gas	0,814*	0,736*	1	0,762*	natural gas	0.202	0.011	1	0.088
nuclear fuels	0,844*	0,892*	0,762*	1	nuclear fuels	0.265	0.217	0.088	1

* significant at the 1%-level

Source: Monthly import prices Germany, 1962 – 2002 (Natural Gas: from 1970, Nucl. Fuels: from 1980 on)

Therefore the introduction of an equilibrium correction term seems appropriate. This leads to the general multivariate model:

$$d(d\mathbf{x}_F) = \mathbf{K}_F\left(\boldsymbol{\mu}_{dF} - d\mathbf{x}_F\right) + \mathbf{B}_F\mathbf{x}_F + \boldsymbol{\Sigma}_F d\mathbf{z}_F \qquad (4\text{-}62)$$

where $\mathbf{K}_F$, $\boldsymbol{\mu}_F$ and $\boldsymbol{\Sigma}_F$ are matrix resp. vector generalizations of their scalar counterparts α_f, μ_f and σ_f. The term $\mathbf{B}_F\,\mathbf{x}_F$ describes the additional tendency of return towards equilibrium price ratios[21]. For a particular fuel f it will read $+b_{ff}x_f + \sum\limits_{\substack{f \\ f' \neq f}} b_{ff'}x_{f'}$. The diagonal elements of $\mathbf{B}_F$ have thus to be negative to obtain a return to the equilibrium, whereas the off-diagonal elements should mostly be positive[22].

Even if symmetry conditions were imposed, the estimation of the entire system would involve more than 30 parameters. More promising is the specification of an a priori restricted system. Thereby the restrictions can be

---

[21] Similar to the error correction term in the error correction models (ECM) of econometric time series analysis.

[22] Formally, the (possibly complex valued) roots of the characteristic equation

$$\det\left|\lambda\begin{bmatrix} \mathbf{I} & 0 \\ 0 & \mathbf{I} \end{bmatrix} - \begin{bmatrix} 0 & \mathbf{I} \\ \mathbf{B}_F & -\mathbf{A}_F \end{bmatrix}\right| = 0$$ determine the stability properties of the overall

equation system (cf. stability analysis in control theory, e.g. Föllinger 1985). If all roots have a negative real part, the system will be stable, i.e. all variables will be stationary (in the econometric sense, cf. e.g. Hamilton 1994). In the case of some roots with zero real part, the system will be indifferent and some or all of the variables x_f will be integrated.

based on the fact that the oil price is still the single most important world energy price. We therefore specify for oil a price model as in Eq. (4-61), taking it to discrete time for estimation purposes:

$$\Delta\Delta x_{Oil} = \kappa_{Oil}\left(\mu_{\Delta Oil} - \Delta x_{Oil}\right) + \sigma_{Oil}\varepsilon_{Oil} \qquad \varepsilon_{Oil} \sim N(0,1) \quad (4\text{-}63)$$

For the other energy carriers, we introduce the deviations from the long-term equilibrium with oil and the oil price change as additional explanatory variables:

$$\Delta\Delta x_f = \kappa_f\left(\mu_{\Delta f} - \Delta x_f\right) + \beta_f\left(x_{Oil} - \theta_f x_f\right) + \gamma_f\Delta\Delta x_{Oil} + \sigma_f\varepsilon_f$$
$$\varepsilon_f \sim N(0,1) \qquad (4\text{-}64)$$

Furthermore the error terms are assumed to be uncorrelated, so that least-squares estimation is unbiased and efficient.

Estimation results are given in Table 4-3. Almost all parameters are significant and the mean-reversion tendency for price changes is with 0.55 to 0.78 rather high, so the price changes have a strong tendency to return to their equilibrium growth rate.

Table 4-3. Estimation results for primary energy prices

	Av. price change $\mu_{\Delta f}$ p.a.		Mean reversion rate κ_f		Error correction β_f		price ratio θ_f		oil price change γ_f		R^2
	coeff.	s.e.	coeff.	s.e.	coeff.	s.e.	coeff.	s.e.	coeff.	s.e.	
crude oil	0.0551	0.0508	0.5512	0.0415							0.276
hard coal	0.0236	0.0044	0.7178	0.0433	0.0236	0.0041	1.948	0.471	0.0855	0.0160	0.426
natural gas	0.0642	0.0293	0.7716	0.0480	0.0552	0.0070	0.902	0.163	0.0154[a]	0.0292	0.424
heavy fuel oil	0.0461	0.0239	0.6140	0.0421	0.0641	0.0191	1.126	0.463	0.6325	0.0596	0.505
nuclear fuels	-0.0129	-0.0308	0.7756	0.0600	0.0177	0.0072	1.169	0.690	0.1515	0.0355	0.438

[a] for gas price changes, the oil price has been lagged by six months in order to account for the traditional natural gas pricing formula, which take into account the lagged oil price
Data source: Monthly import prices Germany, 1962 – 2000,
for natural gas: 1970 – 2000, for nuclear fuels: 1980 – 2000

The error correction term for the price ratio is with 0.018 to 0.064 much lower, hence a return to the equilibrium price ratios is only occurring in a time span of several years. For all fuels except coal, the coefficient of the price ratio is not statistically significantly different from 1. Since the model is formulated in logs, this indicates that there is a constant equilibrium price ratio. For coal the price ratio coefficient is much higher – indicating that the oil price has lower impact on (second order) coal price changes than the coal price level itself. In other words, the coal price is a stationary time series which will only partly follow the oil price movements – a rather plausible result.

It should be noted that the model in its current version does not deal separately with the exchange rate risk and the commodity price risk on the world markets. It is also formulated in nominal prices, leading to rather high price growth rates. Yet it captures key features of the observed price patterns. As measure of model quality, R^2 is reported, since it is readily available from the least-squares regressions. The rather low R^2 should not be interpreted as a sign of poor model fit, since the explained variables are second differences in the logs of prices – which are clearly more difficult to explain than price levels. Also no exogenous explanatory variables have been included, in order to make the model easy to handle for simulations of future prices.

3.2   Electricity prices: fundamental model

Starting from given primary energy prices, this module determines the marginal generation costs as a function of available generation and transmission capacities, primary energy prices, plant characteristics and actual electricity demand. Also the impact of hydro-storage and start-up costs is taken into account. The basic principle of the model is cost minimization in an integrated (in occurrence Europe-wide) power network as described in section 1. Since start-up costs are considered (cf. section 1.4), the objective function is written:

$$G = \sum_{t=1}^{t_{max}} \varphi_t d_t \sum_u \left(C_{Op,u,t} + C_{St,u,t} \right) \tag{4-65}$$

with $C_{Op,u,t}$ the operating costs of unit type u at time t and $C_{St,u,t}$ the corresponding start-up costs. The model is formulated in typical days of one year, in order to be able to describe seasonal effects and seasonal hydro storage as well as start up costs (cf. section 1.6). This is necessary since in the continental European power system hydro power from storage and pump-storage plants is quite of some importance.

t thereby stands for a time segment in a typical day, d_t is the duration of that time segment and φ_t its frequency in the planning horizon (one year). In the current version of the model (partly based on earlier work by Kramer 2002), twelve typical days are distinguished (see Table 4-4), with weekdays and weekends always taken as separate typical days due to their difference in load and price patterns. Furthermore to capture seasonal effects in load, water inflow, sunshine etc., separate typical days are distinguished for every two month period. Within each typical day, time segments of two hours length each are distinguished in order to keep the problem size reasonable.

Table 4-4.   Definition of typical days in the fundamental model within the integrated model

January - February weekday	January- February weekend
March – April weekday	March – April weekend
May – June weekday	May – June weekend
July - August weekday	July - August weekend
September – October weekday	September – October weekend
November – December weekday	November – December weekend

The operating costs are formulated as in Eq. (4-20) and the start-up costs as in (4-21). In order to describe the links between capacity online and output, the Eqs. (4-18) and (4-19) are also included. The demand and supply balance (4-9) is modified to include also pumping energy for hydro storage, leading to:

$$\sum_{u \in U_r} y_{u,t} + \sum_{r'} \left(e_{r',r,t} - e_{r,r',t} \right) \geq D_{r,t} + \sum_{\substack{u \in U_r \\ u \in U_{PS}}} \tilde{y}_{PUMP,u,t} \qquad (4\text{-}66)$$

The regions distinguished within the model are given in Table 4-5. The regions thereby account for the major grid constraints within Europe (cf. e.g. Auer et al. 2002). For the endogenous regions, the full set of equations is specified, for the exogenous regions only the import and export flows to the endogenous regions are specified as time series based on historically observed values.

Table 4-5.   Definition of regions in the fundamental model within the integrated model

Region	Countries	Endogenous/exogenous
Germany	Germany	Endogenous
France	France	Endogenous
Benelux	Belgium, Netherlands, Luxemburg	Endogenous
Alpine Region	Austria, Switzerland	Endogenous
Italy	Italy	Endogenous
Iberian Peninsula	Spain, Portugal	Endogenous
UK and Ireland	United Kingdom, Ireland	Exogenous
Scandinavia	Denmark, Norway, Sweden, Finland	Exogenous
Eastern Europe	Poland, Czechia, Hungary, Slovenia etc.	Exogenous

The capacity restrictions for plants and transmission lines are as described in Eqs. (4-10) and (4-11). For transmission lines, net transfer capacities as reported by ETSO (2003) are taken as basis. A delicate point is the available capacity for hydro storage plants. In fact the aggregation of all hydro plants into one single reservoir oversimplifies the operation restrictions for these plants. Many plants have restrictions on minimum water flow. Restrictions arise also due to hydro cascading and make a full simultaneous use of installed hydro capacity for generation unlikely if not impossible. Unfortunately no detailed information is available on these restrictions at plant level, therefore general assumptions are used: The minimum water flow is set to 10 % of installed hydro generation capacity and an average capacity availability of 0.5 is used for hydro storage plants. Furthermore, the hydro reservoirs in each region are split into two types: small ones for daily storage and large ones for seasonal storage. This is done, since pumping capacities are mostly installed on smaller reservoirs and used for daily or at maximum weekly cycling. If these restrictions are not taken into account, hydro storage release and pumping will lead to very flat price curves in most regions, as a consequence of the peak shaving effect of hydro storage and the valley filling effect of pumping.

Further data sources include UCTE statistics on electricity load and electricity exchanges, a database of power plants throughout Europe and fuel price statistics. Power plants are classified according to the main fuel used and the vintage into 25 classes in order to keep the problem size reasonable (see Table 4-6). The linear programming problem to be solved has nevertheless about 61,000 variables, 83,000 constraints and 220,000 non-zero elements. Implemented in GAMS (cf. Brooke et al. 1998) and using a CPLEX solver (cf. ILOG 2003), computation time on a Pentium III PC is about 3 minutes for a solve from scratch and about 10 s if an initial solution from a previous run is provided.

The model is operated in two different ways: Firstly it is used with historical fuel prices and demand to determine the equilibrium prices for past time periods, which are used as input for estimating the stochastic electricity price model described below. Thereby simulations are carried out for each historical month using the fuel price information available at that time. Secondly the model is operated using Monte-Carlo-simulations of future fuel prices (taken from the model described in section 3.1) as input, in order to derive stochastic equilibrium prices. Due to the possibilities of fuel switches, variations in hydro-use or changes in import-export flows, the obtained relationship between fuel prices and electricity prices is thereby non-linear.

The results of the historical analyses are reported in Figure 4-4, together with the observed prices averaged over all days and hours corresponding to the time segment under consideration.

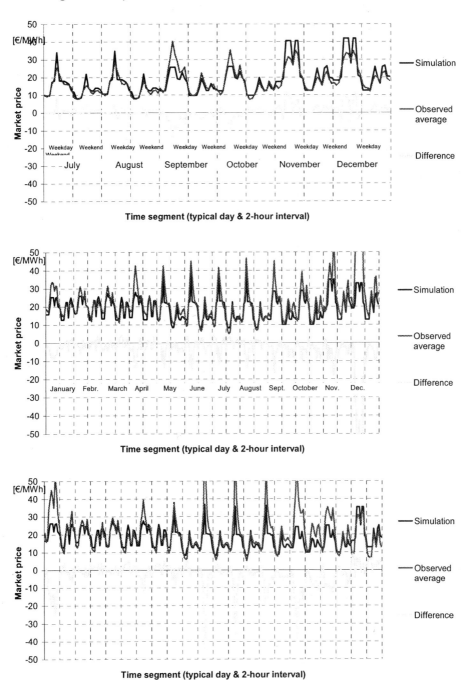

Figure 4-4. Comparison of prices obtained from the fundamental model and historical energy prices

Table 4-6. Definition of power plant types in the fundamental model within the integrated model

Power plant	Fuel	Vintage
coal plant 1	coal	before 1970
coal plant 2	coal	1970 – 1979
coal plant 3	coal	1980 – 1984
coal plant 4	coal	1985 – 1989
coal plant 5	coal	1990 – 1994
coal plant 6	coal	1995 and later
lignite plant 1	lignite	before 1970
lignite plant 2	lignite	1970 – 1979
lignite plant 3	lignite	1980 – 1984
lignite plant 4	lignite	1985 – 1989
lignite plant 5	lignite	1990 – 1994
lignite plant 6	lignite	1995 and later
nuclear plants	nuclear	All
gas fired gas turbine	gas	All
gas fired plants (steam turbines)	gas	All
gas fired combined cycle plant	gas	All
oil fired plants (steam turbines) 1	oil	before 1970
oil fired plants (steam turbines) 2	oil	1970 and later
oil fired gas turbines 1	oil	before 1970
oil fired gas turbines 2	oil	1970 and later
steam-cycle plants with mixed firing 1	oil and gas	before 1970
steam-cycle plants with mixed firing 2	oil and gas	1970 and later
combined cycle plants with mixed firing	oil and gas	All
gas turbines with mixed firing	oil and gas	All
plants with other fuels	waste, biomass etc.	All

It clearly turns out that the model describes well the general price patterns observed, but that prices, even averaged over similar typical days within one month, deviate in some months considerably from those predicted by the fundamental model. This occurs especially in December 2001, when for the first time the (then) price limit of 1000 €/MWh was reached on the German spot market in the evening of 17 December and prices had already been high before and remained rather high until in January. But also in summer 2002, the average midday price observed has been far beyond the one predicted by the model. Partly there are certainly fundamental explications for these deviations from expected average (plant outages in both periods, additionally very low temperatures in December 2001 in whole Europe), but in general it has to be noted that these price peaks occur in periods where the fundamental prices are already rather high – i.e. capacities rather high. In these market phases, the remaining market players with free capacity may charge strategic prices – a phenomenon which has to be taken into account when modeling stochastic price behavior in the third part of the integrated model and which will also be reconsidered when analyzing strategic interaction between players in chapter 5 and long-term investment and price equilibria in chapter 9.

The mean absolute error (MAE) over the whole historical period is 4.26 €/MWh when comparing to the simulated prices to the mean prices in the corresponding time segments. The root mean squared error (RMSE) is with 8.91 €/MWh considerably higher, since it is much more influenced by the empirically observed but fundamentally not well explained peak prices. But since prices beyond 50 €/MWh can hardly be explained by any model based on marginal generation costs, even if start-up costs are accounted for, these prices should not be considered when measuring the model quality of the fundamental model. Measured on this restricted data set, the MAE decreases to 3.48 €/MWh, whereas the RMSE drops to 5.08 €/MWh.

3.3   Electricity prices: stochastic model

The fundamental model described in the previous section provides as shown a rather realistic picture of the expected generation costs and corresponding market prices. Besides these deterministic impacts from marginal generation costs, time-varying demand, capacity constraints etc., electricity prices are however also subject to stochastic changes – due e. g. to load fluctuations, unforeseen outages or trader speculation. These stochastic fluctuations are modeled in the integrated model based on a stochastic modeling approach as described in section 2. The model is specified as a mean-reversion model given the empirical observations and the results of earlier research (see above). However one expects that the prices do not tend to return to a time-invariant mean, but rather to one which is determined by the fundamentals governing the electricity market. Therefore, the following model is specified:

$$\Delta \ln p_{E,d,h} = \kappa_{E,h} \left[\ln p_{G,d-1,h} - \ln p_{E,d-1,h} + \beta_{E,h} + \beta_{K,h} \left(K_{PL,fr,d,h} - \overline{K}_{PL,fr} \right) + \right.$$

$$\left. \sum_{wd' \in \{1,2,4,5\}} \left(\chi_{E,wd',h} 1_{wd(d)=wd'} \right) + \chi_{E,hd,h} 1_{hd(d)} + \chi_{E,sd,h} 1_{sd(d)} \right] + \varepsilon_{E,d,h}$$

$$\varepsilon_{E,d} \sim N\left(0, < \sigma_{E,d} > \Omega < \sigma_{E,d} > \right)$$

$$(4\text{-}67)$$

that is the price change is dependent on the logarithmic difference between the equilibrium price $p_{G,h,t}$ and the actual price $p_{E,t}$ adjusted possibly by a term depending on the free capacity $K_{PL,fr}$. Additionally the parameters $\chi_{E,wd',h}$ associated with the weekday dummy variables $1_{wd(d)=wd'}$ are introduced to cope with possible systematic price variation between weekdays. Similarly the parameters $\chi_{E,hd,h}$ and $\chi_{E,sd,h}$ are used to describe the effects of holidays hd within the week (such as Easter Monday) and of special days sd which are not public holidays but have reduced activity levels such as bridging days between holidays and the weekend or the days between Christmas and New Year's day.

One should note that albeit the model is formulated as a mean-reversion model, the obtained time series are not necessarily stationary. This is due to the fact that the equilibrium prices are influenced by the fuel prices, the stochastics of which are described through a separate model (cf. section 3.1). Since these are not all stationary time series, also the equilibrium prices and the stochastic electricity prices are not necessarily stationary.

Given the evidence provided by Cuaresma et al. (2003), each hour is modeled as a separate stochastic process similarly to the approach adopted by Ramanathan et. al (1997). Additionally the weekend prices are also modeled as separate stochastic processes since the fundamentals, notably the load characteristics, are rather different on weekends than during the week. In most markets, trading is furthermore strongly reduced on weekends. Consequently prices for electricity delivered on weekends as well as on Mondays are fixed on power exchanges like the German EEX already on Friday. This leads to the following, slightly modified model for prices on weekends:

$$\Delta \ln p_{E,d',h} = \kappa_{E,h} \left(\ln p_{G,d'-1,h} - \ln p_{E,d'-1,h} + \beta_{E,h} + \beta_{K,h} \left(K_{PL,fr,d',h} - \overline{K}_{PL,fr} \right) \right.$$
$$\left. + \chi_{E,7,h} 1_{wd(d')=7} + \chi_{E,Fr,h} \ln p_{E,Fr(d'),h} + \chi_{E,hd,h} 1_{hd(d')} \right) + \varepsilon_{E,d',h}$$
$$\varepsilon_{E,d'} \sim N\left(0, <\sigma_{E,d'} > \Omega < \sigma_{E,d'} >\right)$$

$$(4\text{-}68)$$

Here the day index d' includes only Saturdays and Sundays, and an additionally explanatory term $\chi_{E,Fr(d'),h}$ describes the effect of the prices on the Friday $Fr(d')$ preceding the day d' on the current price level. Hence weekend prices depend both on the prices observed during the weekend before and the last prices recorded during the week.

The entire model is formulated in logs of prices, on the one hand to prevent negative prices, so far not observed in Europe, on the other hand to describe at least approximately the frequency of price spikes.

For the specification of the stochastic term $\varepsilon_{E,d,h}$ the following points have been retained: Firstly volatility clustering is important in electricity prices, therefore a GARCH-model specification is chosen. Secondly, part of the stochastic price fluctuations is due to variations in the load or available capacity. Yet, the impact of these fluctuations depends on the slope of the merit order curve (see Figure 4-5). Therefore the change $\Lambda_{E,d,h}$ in marginal costs for a change in load by ±15 percent is included as explanatory variable in the GARCH-model of volatility:

$$\sigma_{E,d,h}^{2} = \gamma_{E,0,h} + \gamma_{E,1,h}\varepsilon_{E,d-1,h}^{2} + \gamma_{E,2,h}\sigma_{E,d-1,h}^{2} + \gamma_{E,3,h}\left(\Lambda_{E,d,h}^{'} - \overline{\Lambda}_{E,h}\right) \quad (4\text{-}69)$$

An analogous specification is used for the volatility on weekends.

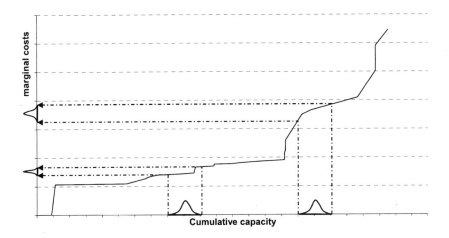

Figure 4-5. Merit order curve and impact of load variations (or variations in available capacities) on marginal costs

As alternatives, a pure GARCH-model is tested as well as a specification, where the variance is dependent on the price level of the previous day. The latter models the fact that volatility tends to increase in electricity markets with price levels and is formally written:

$$\sigma_{E,d,h}^{2} = \gamma_{E,0,h} + \gamma_{E,1,h}\varepsilon_{E,d-1,h}^{2} + \gamma_{E,2,h}\sigma_{E,d-1,h}^{2} + \gamma_{E,3,h}\left(\ln p_{E,d,h} - \overline{\ln p_{E,h}}\right)$$
$$(4\text{-}70)$$

In principle, also specifications of the stochastic term with non-normal distributions like Markov regime switching models (cf. section 2.3.4) or the transformed diffusion model of Barlow (2002) (cf. section 2.3.5) would be interesting candidates to describe even better the observed price spikes. But these require additional parameters and have therefore at first discarded in order to avoid too complicated specifications.

A further point, which is taken into account in the stochastic specification, is the correlation between the stochastic terms of different hours. This is crucial, if realistic price simulations are to be obtained. Among the multitude of possible multivariate GARCH specifications (cf. Engle, Mezrich 1996), a specification of medium complexity has been retained, namely the constant conditional correlation model. It is described through the relationships (already contained in Eqs. (4-67) and (4-68)):

$$E\left[\varepsilon_{E,d}\varepsilon_{E,d}{}^{T}\right] = <\sigma_{E,d}> \Omega <\sigma_{E,d}>$$
$$\Leftrightarrow \quad \Omega = <\sigma_{E,d}>^{-1} E\left[\varepsilon_{E,d}\varepsilon_{E,d}{}^{T}\right] <\sigma_{E,d}>^{-1}$$
$$E\left[\varepsilon_{E,d'}\varepsilon_{E,d'}{}^{T}\right] = <\sigma_{E,d'}> \Omega' <\sigma_{E,d'}>$$
$$\Leftrightarrow \quad \Omega' = <\sigma_{E,d'}>^{-1} E\left[\varepsilon_{E,d'}\varepsilon_{E,d'}{}^{T}\right] <\sigma_{E,d'}>^{-1}$$

$$(4\text{-}71)$$

Here $<\sigma_{E,d}>$ and $<\sigma_{E,d'}>$ are diagonal matrices containing the hourly standard deviations, whereas Ω resp. Ω' are correlation matrices between the volatility corrected stochastic terms of the different hours. In order to estimate them, first the Eqs. (4-67) and ((4-68) may be estimated separately hour by hour together with the corresponding volatility equation. Then the correlation matrix can be computed by inserting the obtained residuals into the right halves of Eq. (4-71). A so-called seemingly-unrelated-regressions- or SUR-estimator (cf. e.g. Greene (2000)), which simultaneously estimates the parameters for all hourly equations and the correlation matrix, may theoretically be more efficient, yet it is also less robust against misspecifications of the error term.

The estimation results are summarized in Table 4-7 to Table 4-10. The mean-reversion parameters are throughout positive and significant (cf. Table 4-7 and Table 4-8). The excess price factor has significant positive values mostly during peak load hours within the week. This corresponds to the expectation from the fundamental model, that market power will be mostly relevant when load is high. But during many of these hours the impact of additional free capacities on prices is significantly positive, instead of being negative as one would expect from the market power argument. On weekdays the free capacity has the expected negative impact on prices only during the hours 7 and 8 in the morning and 19 to 21 in the evening. The inverse effect can partly be explained through imperfections in the fundamental model. As can be seen in Figure 4-5, the fundamental model predicts a very sharp price spike in summer at noon. For the preceding and the subsequent hours the observed prices are almost every time above the expected ones. In winter, when free capacities are lower, this deviation is less observed. Consequently the statistical model will impute the higher offset from equilibrium prices to increased free capacities – whereas in reality it is due e.g. to limited peak shaving capabilities of hydro power.

As far as systematic differences between weekdays are concerned, the prices are significantly lower during the night from Sunday to Monday but higher during daytime on Monday (cf. Table 4-7). This corresponds to differences in load, which rises only on Monday morning from weekend levels to weekday level or even above, e.g. through additional start-up consumption.

Table 4-7. Estimation results for the stochastic model of electricity prices – regression coefficients for weekdays

	Mean reversion param. $\kappa_{E,h}$		Excess price factor $\beta_{E,h}$		Impact free capacity $\beta_{K,h}$		Weekday dummies $\chi_{E,wd,h}$ Monday		Tuesday		Thursday		Friday		Holiday dummy $\chi_{E,hd,h}$		Special day dummy $\chi_{E,hd,h}$	
	coeff.	a.s.e	coeff.	a.s.e.	coeff.	a.s.e	coeff.	a.s.e.	coeff.	a.s.e.	coeff.	a.s.e.	coeff.	a.s.e	coeff.	a.s.e.	coeff.	a.s.e.
1	**0.37**	0.03	**0.18**	0.06	0.47	0.28	**-0.49**	0.08	**0.36**	0.08	0.04	0.07	0.03	0.07	**-0.41**	0.09	**-0.13**	0.01
2	**0.37**	0.04	0.07	0.06	0.12	0.36	**-0.47**	0.08	**0.28**	0.09	0.02	0.09	-0.02	0.08	**-0.66**	0.10	-0.01	0.39
3	**0.38**	0.03	0.01	0.06	-0.11	0.39	**-0.46**	0.08	**0.25**	0.08	-0.07	0.09	-0.09	0.08	**-0.81**	0.10	-0.06	0.15
4	**0.35**	0.03	-0.02	0.06	-0.15	0.38	**-0.44**	0.08	**0.28**	0.08	-0.07	0.08	-0.09	0.08	**-0.91**	0.13	-0.10	0.08
5	**0.40**	0.03	0.03	0.06	-0.06	0.34	**-0.51**	0.08	**0.28**	0.08	-0.04	-0.08	-0.05	0.07	**-0.96**	0.10	**-0.16**	0.02
6	**0.43**	0.03	**0.16**	0.06	0.18	0.40	**-0.34**	0.07	**0.25**	0.07	0.01	0.07	0.05	0.07	**-1.64**	0.13	**-0.14**	0.02
7	**0.58**	0.02	**-0.09**	0.04	**-0.85**	0.29	**-0.11**	0.04	0.05	0.04	-0.07	0.04	-0.04	0.05	**-1.68**	0.06	**-0.18**	0.01
8	**0.50**	0.02	**0.12**	0.04	**-1.58**	0.30	0.00	0.04	0.10	0.05	0.03	0.05	0.05	0.04	**-2.15**	0.09	**-0.10**	0.03
9	**0.48**	0.03	**0.19**	0.05	**1.54**	0.46	0.07	0.06	0.06	0.06	0.02	0.07	0.08	0.06	**-2.41**	0.15	0.05	0.13
10	**0.51**	0.03	**0.37**	0.05·	**3.03**	0.50	0.05	0.06	0.02	0.06	-0.01	0.06	-0.03	0.06	**-1.69**	0.10	**-0.15**	0.02
11	**0.30**	0.03	0.04	0.08	**-0.69**	0.78	0.04	0.09	0.10	0.10	0.01	0.09	-0.10	0.09	**-2.01**	0.21	**-0.17**	0.07
12	**0.32**	0.03	**0.26**	0.09	**1.84**	0.85	**0.33**	0.11	**0.23**	0.11	0.07	0.10	-0.03	0.11	**-1.66**	0.23	**-0.28**	0.01
13	**0.39**	0.03	**0.48**	0.07	**3.76**	0.65	**0.20**	0.08	-0.03	-0.07	-0.06	0.08	**-0.23**	0.09	**-1.28**	0.14	-0.06	0.37
14	**0.48**	0.03	**0.44**	0.06	**3.46**	0.50	**0.13**	0.06	-0.03	-0.06	-0.07	0.06	**-0.30**	0.07	**-1.48**	0.10	-0.03	0.58
15	**0.52**	0.03	**0.37**	0.05	**1.55**	0.43	**0.17**	0.06	0.02	0.06	-0.06	0.06	**-0.34**	0.06	**-1.42**	0.09	-0.11	0.22
16	**0.44**	0.03	**0.22**	0.04	**0.67**	0.39	**0.21**	0.06	0.04	0.06	-0.02	0.06	**-0.37**	0.07	**-1.53**	0.10	-0.08	0.36
17	**0.41**	0.03	**0.13**	0.04	**1.79**	0.25	**0.21**	0.06	**0.11**	0.06	0.04	0.06	**-0.33**	0.06	**-1.46**	0.08	-0.10	0.26
18	**0.46**	0.03	0.00	0.03	0.13	0.26	**0.13**	0.05	**0.15**	0.05	0.08	0.05	**-0.14**	0.04	**-1.19**	0.07	**-0.20**	0.03
19	**0.34**	0.03	-0.04	0.04	**-1.12**	0.26	**0.21**	0.06	**0.15**	0.06	**0.14**	0.06	**-0.15**	0.05	**-1.35**	0.10	**-0.32**	0.03
20	**0.35**	0.04	-0.04	0.04	**-0.67**	0.24	**0.21**	0.07	**0.11**	0.06	**0.12**	0.06	**-0.18**	0.06	**-1.06**	0.12	**-0.25**	0.05
21	**0.46**	0.04	0.01	0.03	**-1.24**	0.20	**0.16**	0.03	**0.11**	0.04	**0.09**	0.04	**-0.10**	0.03	**-0.57**	0.07	-0.10	0.05
22	**0.23**	0.03	-0.03	0.05	**-0.48**	0.39	**0.22**	0.08	**0.15**	0.08	0.16	0.08	**-0.16**	0.07	**-0.29**	0.15	-0.09	0.19
23	**0.19**	0.03	**0.17**	0.07	**1.29**	0.39	0.04	0.09	0.14	0.09	0.10	0.09	0.00	0.08	**-0.32**	0.12	**-0.11**	0.00
24	**0.20**	0.03	0.00	0.07	**0.91**	0.33	-0.11	-0.08	0.18	0.09	0.10	0.09	**0.24**	0.09	**-0.35**	0.14	-0.10	0.23

Coefficients significant at the 5 % level are marked in **bold**. Testing is thereby done using the asymptotic standard errors. Source: Own calculations using observations between July 1,2000 and October 1,2002

Tuesday is from an electricity price point of view also systematically different from Wednesday which is taken as reference. Both during the night hours and in the evening, prices are significantly higher, which might be due to higher load or to trading effects e.g. that forced outages during the weekend or unplanned prolongations of down-times which appear on Monday morning will impact prices only from Tuesday onwards. Whereas Thursday is very similar to Wednesday, Friday has clearly a lower load and consequently also lower prices from 1 p.m. onwards.

Table 4-8. Estimation results for the stochastic model of electricity prices – regression coefficients for weekends

	Mean reversion param. $\kappa_{E,h}$		Excess price factor $\beta_{E,h}$		Impact free capacity $\beta_{K,h}$		Weekday dummy Saturday $\chi_{E,wd,h}$		Impact Friday $\chi_{E,hd,h}$	
	coeff.	a.s.e.	coeff.	a.s.e.	coeff.	a.s.e.	coeff.	a.s.e.	coeff.	a.s.e.
1	**0.40**	0.06	**-0.45**	-0.15	-0.94	-0.69	**1.16**	0.24	**0.51**	0.16
2	**0.43**	0.06	**-0.50**	-0.15	-1.16	-0.73	**1.17**	0.25	**0.45**	0.14
3	**0.84**	0.05	-0.11	-0.07	-1.20	-0.73	**0.50**	0.08	**0.50**	0.08
4	**1.04**	0.06	0.00	-0.06	-1.48	-0.34	**0.31**	0.06	**0.58**	0.05
5	**0.59**	0.11	-0.21	-0.14	**-2.44**	-0.76	**0.74**	0.26	**0.85**	0.14
6	**0.94**	0.06	0.08	0.09	**-2.74**	-0.62	**0.30**	0.09	**0.63**	0.08
7	**0.73**	0.09	-0.47	-0.28	-0.02	-1.27	**0.86**	0.23	**0.65**	0.12
8	**0.83**	0.05	**-1.47**	-0.51	2.84	2.20	**1.40**	0.23	**0.41**	0.11
9	**0.62**	0.06	**-0.96**	-0.34	1.23	1.43	**1.18**	0.23	**0.28**	0.07
10	**0.56**	0.08	**-0.82**	-0.26	1.88	1.14	**1.16**	0.25	**0.32**	0.07
11	**0.68**	0.07	**-0.61**	-0.15	0.97	0.74	**0.69**	0.13	**0.31**	0.04
12	**0.55**	0.06	-0.30	-0.19	-0.45	-0.97	**0.63**	0.14	**0.22**	0.04
13	**0.36**	0.06	**-0.77**	-0.30	2.00	1.50	**1.08**	0.24	**0.33**	0.12
14	**0.40**	0.06	**-0.81**	-0.23	1.70	1.02	**1.04**	0.20	**0.22**	0.10
15	**0.55**	0.07	0.05	0.17	-1.33	-0.76	**0.61**	0.14	**0.35**	0.06
16	**0.60**	0.07	0.18	0.17	**-2.12**	-0.79	**0.54**	0.13	**0.41**	0.08
17	**0.44**	0.06	**-0.76**	-0.19	**1.66**	0.77	**0.95**	0.20	**0.57**	0.09
18	**0.44**	0.05	**-0.38**	-0.12	0.38	0.51	**0.81**	0.14	**0.52**	0.09
19	**0.45**	0.05	-0.20	-0.10	-0.07	-0.47	**0.47**	0.10	**0.52**	0.09
20	**0.50**	0.05	-0.05	-0.08	0.18	0.42	**0.19**	0.06	**0.62**	0.09
21	**0.37**	0.04	-0.14	-0.12	1.05	0.84	0.07	0.07	**0.73**	0.12
22	**0.39**	0.05	**-0.23**	-0.06	**2.50**	0.60	-0.05	-0.05	**0.58**	0.15
23	**0.40**	0.06	0.04	0.05	0.82	0.53	**-0.16**	-0.05	**0.42**	0.10
24	**0.53**	0.06	**-0.17**	-0.06	-0.20	-0.53	**0.13**	0.05	**0.70**	0.12

Coefficients significant at the 5 % level are marked in **bold**. Testing is thereby done using the asymptotic standard errors. Source: Own calculations using observations between July 1, 2000 and October 1, 2002

For public holidays, a strong and systematic impact on loads and prices is observed, whereas the effect of the special days is much less pronounced and only partly significant. On weekends a systematic effect of the price on the preceding Friday is found (cf. Table 4-8). Moreover, prices on Saturdays are throughout significantly higher than on Sundays, except for the late evening.

The volatility modeling shows clear effects of volatility clustering with both past volatilities and past price changes having significant effects on weekdays (cf. Table 4-9), whereas the evidence is less clear cut on weekends. This could be due to the smaller data base or/and to the alternation of Saturdays and Sundays, which differ considerably in their characteristics.

Table 4-9. Estimation results for the stochastic model of electricity prices – volatility coefficients and likelihood values for weekdays

	Volatility coefficients								**Log-Likelihood value**		
Constant $\gamma_{E,0,h}$		Past changes $\gamma_{E,1,h}$		Past volatility $\gamma_{E,2,h}$		Grad. merit order $\gamma_{E,3,h}$		Model	Simple GARCH	GARCH with price
coeff.	a.s.e.	coeff.	a.s.e.	coeff.	a.s.e.	coeff.	a.s.e.			
1	**0.004**	0.0012	**0.124**	0.0416	**0.745**	0.0742	-0.0003	0.0004	253.5	253.5	253.2
2	**0.003**	0.0006	**0.149**	0.0262	**0.816**	0.0273	0.0008	0.0005	93.8	92.7	94.0
3	**0.001**	0.0004	**0.111**	0.0162	**0.869**	0.0166	-0.0001	0.0003	**39.9**	36.2	**40.2**
4	**0.002**	0.0005	**0.148**	0.0187	**0.832**	0.0168	0.0004	0.0006	14.5	12.0	14.5
5	**0.002**	0.0005	**0.149**	0.0225	**0.831**	0.0212	0.0001	0.0005	43.6	40.7	43.7
6	**0.005**	0.0010	**0.154**	0.0411	**0.763**	0.0450	**0.0022**	0.0007	**100.1**	95.2	95.2
7	**0.004**	0.0008	**0.282**	0.0394	**0.698**	0.0363	**-0.0012**	0.0004	**88.8**	84.7	**87.6**
8	**0.009**	0.0021	**0.490**	0.0838	**0.465**	0.0728	-0.0043	0.0063	51.5	50.9	51.8
9	**0.016**	0.0028	**0.490**	0.0797	**0.450**	0.0685	**0.0340**	0.0074	**-35.9**	-43.8	**-40.7**
10	**0.009**	0.0025	**0.134**	0.0219	**0.741**	0.0500	0.0033	0.0032	-2.9	-4.5	**-1.9**
11	**0.018**	0.0033	**0.399**	0.0649	**0.409**	0.0660	0.0047	0.0195	-11.1	-11.7	**-8.7**
12	**0.005**	0.0016	**0.200**	0.0379	**0.779**	0.0385	**0.0088**	0.0044	**-135.2**	-138.2	**-132.5**
13	**0.006**	0.0013	**0.164**	0.0239	**0.764**	0.0318	-0.0065	0.0052	-1.6	-3.4	**0.6**
14	**0.008**	0.0014	**0.182**	0.0324	**0.720**	0.0378	-0.0023	0.0032	-3.0	-4.8	**-2.8**
15	**0.010**	0.0030	**0.149**	0.0333	**0.679**	0.0747	-0.0090	0.0065	27.8	26.0	**29.2**
16	**0.017**	0.0032	**0.325**	0.0634	**0.364**	0.0971	**0.0344**	0.0095	**106.4**	99.9	100.2
17	**0.007**	0.0010	**0.317**	0.0701	**0.579**	0.0561	**0.0017**	0.0005	**150.7**	146.1	146.7
18	**0.004**	0.0005	**0.302**	0.0452	**0.678**	0.0333	**0.0008**	0.0002	**129.5**	125.6	**128.4**
19	**0.004**	0.0008	**0.267**	0.0453	**0.676**	0.0385	**0.0015**	0.0003	**172.7**	160.8	**177.7**
20	**0.004**	0.0012	**0.202**	0.0489	**0.709**	0.0637	**0.0014**	0.0005	**185.7**	175.9	**190.4**
21	**0.004**	0.0010	**0.469**	0.0427	**0.511**	0.0419	**0.0010**	0.0004	**233.0**	228.9	228.9
22	**0.013**	0.0018	**0.424**	0.0675	0.048	0.0827	**0.0020**	0.0007	**315.9**	313.1	**323.1**
23	**0.003**	0.0010	**0.225**	0.0597	**0.584**	0.0949	-0.0003	0.0003	389.9	389.3	**392.6**
24	**0.005**	0.0010	**0.306**	0.0630	**0.475**	0.0759	-0.0009	0.0005	362.4	361.2	362.1

Coefficients significant at the 5 % level are marked in **bold**. Testing is thereby done using the asymptotic standard errors

For the log-likelihoods a likelihood ratio test using a χ^2-statistic with 1 degree of freedom is used, significant differences are again marked in **bold**.

Source: Own calculations using observations between July 1, 2000 and October 1, 2002

The impact of the merit order gradient turns out to be significant in eleven hours out of twenty-four on weekdays (cf. Table 4-9), whereas on weekends there are eighteen significant coefficients (cf. Table 4-10). But even when the coefficient is not significant, the improvement in log-likelihood can be (and conversely), so that a total of thirty-one significant enhancements through the inclusion of the gradient term can be found. For the alternative specification with the price level explaining volatility, significant improvements are obtained twenty-nine times. When comparing directly the log-likelihoods, the chosen specification outperforms the standard model eleven times by more than the critical value of the χ^2-statistic.

Table 4-10. Estimation results for the stochastic model of electricity prices – volatility coefficients and likelihood values for weekends

	Volatility coefficients								Log-Likelihood value		
	Constant $\gamma_{E,0,h}$		Past changes $\gamma_{E,1,h}$		Past volatility $\gamma_{E,2,h}$		Grad. merit order $\gamma_{E,3,h}$		Model	Simple GARCH	GARCH with price
	coeff.	a.s.e.	coeff.	a.s.e.	coeff.	a.s.e.	coeff.	a.s.e.			
1	**0.039**	0.0142	0.091	0.0884	0.131	0.3070	**-0.0127**	0.0038	24.0	23.2	23.7
2	**0.049**	0.0099	**0.231**	0.1146	0.156	0.1426	**-0.0300**	0.0072	**-17.4**	-24.9	**-20.5**
3	**0.031**	0.0061	**0.645**	0.1173	**0.335**	0.0758	**-0.0112**	0.0053	-83.1	-84.1	**-67.4**
4	**0.076**	0.0091	**0.980**	0.1164	0.000	0.0147	**-0.0294**	0.0082	**-115.7**	-120.0	**-105.2**
5	**0.075**	0.0169	**0.686**	0.1143	**0.294**	0.0818	**-0.0373**	0.0121	**-172.8**	-175.7	**-157.6**
6	**0.295**	0.0232	**0.980**	0.1312	0.000	0.0089	**-0.2469**	0.0249	**-212.1**	-230.5	**-203.9**
7	**0.006**	0.0015	**0.065**	0.0151	**0.915**	0.0144	**0.0089**	0.0033	**-203.9**	-244.2	**-193.4**
8	**0.007**	0.0032	**0.207**	0.0432	**0.773**	0.0389	**0,0083**	0.0036	**-117.7**	-123.6	-123.8
9	**0.015**	0.0075	0.078	0.0453	**0.685**	0.1424	**0.0173**	0.0077	1.3	-5.1	**-2.0**
10	**0.030**	0.0084	**0.254**	0.0898	0.013	0.2035	0.0070	0.0054	50.7	50.2	53.3
11	**0.024**	0.0102	0.143	0.0869	0.000	0.3584	**-0.0110**	0.0043	**92.1**	88.4	**90.5**
12	**0.028**	0.0075	0.125	0.0773	0.000	0.2773	**-0.0151**	0.0048	**80.2**	76.6	76.7
13	**0.001**	0.0002	**0.058**	0.0000	**0.921**	0.0000	**0.0045**	0.0001	**92.9**	73.8	**79.3**
14	**0.001**	0.0005	0.001	0.0245	**0.964**	0.0240	**0.0026**	0.0012	**80.8**	77.9	79.7
15	**0.019**	0.0050	**0.370**	0.1027	0.000	0.1639	-0.0182	0.0128	89.8	88.6	88.7
16	**0.038**	0.0053	**0.201**	0.0635	0.000	0.0956	**-0.1148**	0.0332	**51.1**	47.0	**50.9**
17	**0.032**	0.0060	**0.353**	0.1025	0.000	0.1254	**-0.0378**	0.0178	**44.5**	40.3	40.4
18	**0.032**	0.0104	**0.303**	0.1108	0.000	0.2537	0.0082	0.0059	42.0	40.7	41.2
19	**0.020**	0.0050	**0.265**	0.0903	0.179	0.1406	**-0.0157**	0.0038	**76.5**	67.2	68.6
20	**0.015**	0.0069	0.192	0.1057	0.378	0.2128	-0.0084	0.0045	**73.7**	70.3	**73.0**
21	**0.030**	0.0024	0.000	0.0000	0.000	0.0000	**-0.0106**	0.0046	90.3	90.3	90.3
22	**0.001**	0.0005	0.091	0.0483	**0.889**	0.0400	**0.0018**	0.0007	**113.2**	109.3	109.3
23	**0.001**	0.0004	**0.073**	0.0329	**0.894**	0.0357	0.0008	0.0005	**113.6**	104.5	**112.8**
24	**0.030**	0.0125	0.197	0.1195	0.016	0.3802	0.0009	0.0078	57.4	56.7	56.9

Coefficients significant at the 5 % level are marked in **bold**. Testing is thereby done using the asymptotic standard errors

For the log-likelihoods a likelihood ratio test using a χ^2-statistic with 1 degree of freedom is used, significant differences are again marked in **bold**.

Source: Own calculations using observations between July 1, 2000 and October 1, 2002

But also the price dependent specification outperforms eleven times the gradient dependent one[23]. So both models seem of similar statistical quality, the advantage of the chosen specification being that it is based on a fundamental explanation.

4. APPLICATION

In order to use the integrated model as a tool for the simulation of future electricity prices, the different modules have been implemented as software tools using mainly GAMS and MATLAB as software platforms. Figure 4-6 gives an overview of the implementation of the different components of the integrated electricity model. All data are stored in a common data base and then delivered to the different modules. The system is highly automatized, so that both ex-post and ex-ante calculations can be done rapidly.

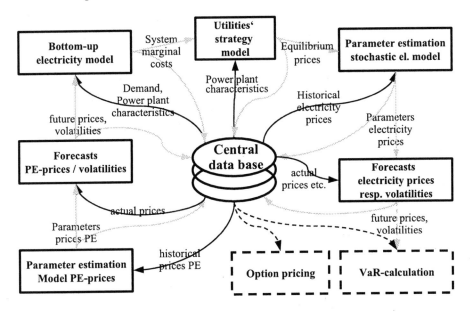

In Figure 4-7, examples of price simulations are shown, which have been determined using the integrated model. In that case, the data base for the

---

[23] Note that the likelihood-ratio test is not exact in this case since both models have the same number of degrees of freedom. For an exact test, a non-nested model test has to be employed. But this will hardly change the general result.

model has consisted of the prices until end of 1 October 2002. On that basis, prices for November 2002 have been simulated. The real market prices for the second and the third week in November have been also included in the figure. Obviously the shape and level of the simulated prices are in general similar to the observed ones although of course the exact price pattern is not matched. Two peaks of varying height appear notably around noon and in the evening at about 6 p.m. whereas prices during night time are rather low.

A comparison of the price ranges implied by different price models yields the results shown in Figure 4-8. Here prices in August 2002 have been forecasted based on information available at end of June 2002. It is evident that modeling electricity price developments based on Wiener processes gives rise to extremely high uncertainty ranges in the medium and long run. Mean reversion processes provide more realistic estimates of price ranges and this is also true for the integrated model. Of course, the price range observed in reality is even smaller, since it does not contain all possible realizations of the stochastic price processes. An important difference between the simple mean-reversion model and the integrated model is that the uncertainty range in the integrated model increases with increasing forecasting horizon due to the impact of uncertain primary energy prices. The mean-reversion model on the contrary, shows no increase after about one week.

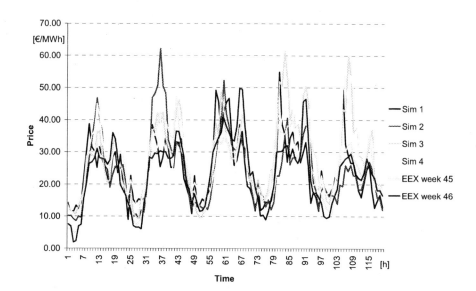

Figure 4-7. Comparison of simulated (as from 1 October 2002) and observed spot market prices in November 2002

Figure 4-9 shows a comparison of price forecasts and uncertainty ranges implied by the integrated model with forward prices and real prices for December 2002. Here, the average as well as the 10 % and the 90 % quantile of the simulations is compared to the actual average spot prices and the forward prices available at the time of simulation and just before the product started to be delivered. In that particular case, the integrated model provides a better forecast for the peak price than the forward price does, both five and one month in advance. Also the uncertainty range looks plausible. But the base forecast is not very good – in fact the observed price is here below the 10 %-quantile of the forecast and also the forward market performed slightly better.

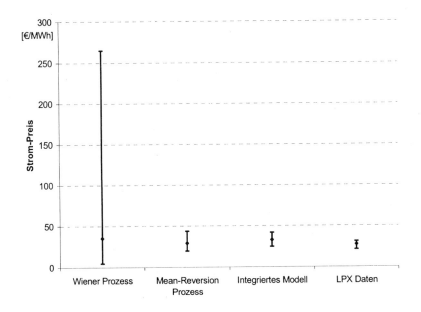

Figure 4-8. Expected price variations (10 % quantile and 90 % quantile) for hour 9 on weekdays in August 2002 as from 30 June 2002

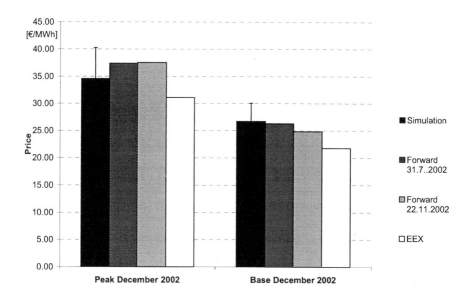

Figure 4-9. Comparison of price forecasts of the integrated model (from 30 June 2001), forward prices and actual prices for standard products November 2001

A more systematic comparison is provided by Figure 4-10. Here the results of ex-post simulation of the prices are compared to actual prices. Thereby, the average of the price simulations obtained at the end of the month preceding delivery is compared to the actual spot market outcomes and to the corresponding forward quotes. Obviously, the simulation provides results which do not align on the forward prices. But the average error, as measured by the mean absolute error (MAE), is 3.6 €/MWh for the simulation model and 4.1 €/MWh for the forward prices. So the simulation model performs on average slightly better then the market does, even if the difference is hardly significant. In terms of the root mean square error (RMSE), the result is similar: The simulation model has an RMSE of 5.8 €/MWh, the forward prices of 7.1 €/MWh.

Figure 4-11 shows for the ex-post simulations how the price uncertainty, measured as the difference between the 10 % and the 90 % quantile of the simulation results, raises as the time to delivery increases. This increase is not very strong and also sometimes decreases occur, given the differences in volatility of the various months. But clearly the model neither provides extreme uncertainty ranges for longer time horizons nor is the price uncertainty limited by an upper bound.

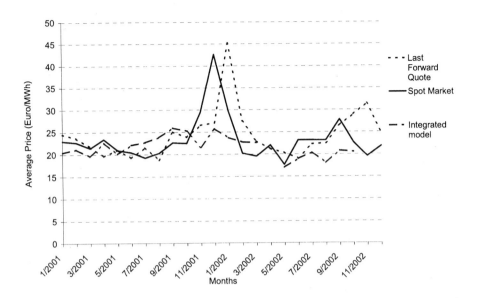

Figure 4-10. Comparison of price forecasts of the integrated model, last forward quotes and actual prices for monthly base products 2001 and 2002

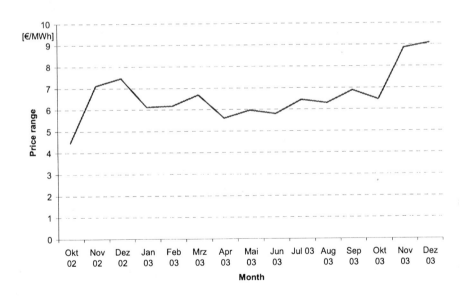

Figure 4-11. Comparison of price forecasts of the integrated model, last forward quotes and actual prices for monthly base products 2001 and 2002

In general, the analysis carried out has shown that a combination of fundamental analysis and stochastic modeling provides an approach to price simulation which takes into account the technical restrictions governing electricity production and transmission as well as the effects of new information arrival and other stochastic effects. Also the dependency of electricity prices on the fuel prices is included in this approach and it can be used to derive both price forecasts and uncertainty ranges for the future development of prices. These can be used both for the operational management of generation and trading portfolios and for assessing the risks associated with these portfolios. This will be shown in chapter 6 and chapter 7. Chapter 9 will develop complementary approaches, which allow tackling price uncertainty in the longer run, when the assumption of a known generation park is not justified. Before, we will now turn to the impact of competitors strategies on electricity prices and own strategies.

Chapter 5

MODELING COMPETITION IN THE ELECTRICITY INDUSTRY

Besides price uncertainty, the strategies of the competitors also affect the decisions of electric utilities. Models of competitive electricity markets have to date mostly been developed to analyze longer term equilibria on the wholesale market. These models are briefly reviewed in section 1. Section 2 takes then a different view on competition in electricity markets, emphasizing the role of retail competition. Here a model developed by the author is discussed, treating electricity delivered by different producers as heterogeneous products. This model is then applied in section 3 to the German power market.

1.   COMPETITION ON THE WHOLESALE MARKET: COURNOT-NASH COMPETITION AND COMPETITION WITH BID CURVES

In general two types of approaches have been used to model the competition on the whole sale market. The first one is using the Cournot-Nash framework (cf. e. g. Andersson / Bergmann, 1995; Borenstein et al., 1999; Ellersdorfer et al., 2001, Ellersdorfer et al., 2003; see also the overview in Smeers, 1997). Underlying are the assumptions of electricity as a homogenous good and of market equilibria being determined through the capacity setting decisions of suppliers. This model type is however only appropriate for the description of the medium to long-term equilibrium determination. Namely the existence of a Nash equilibrium in this modeling framework requires a substantial negative own price elasticity for electricity. While most empirical studies (e. g. Dennerlein 1990, Dahl 1994) agree that significant negative own price elasticities exist for electricity in the longer run, for the short run both empirical studies and most practitioners agree on the electricity price elasticity to be negligible or even non-existent.

The second one is modeling the price equilibria on the wholesale market as equilibria of firms bidding with supply (and possibly demand curves) into the wholesale market. The first models using this approach were developed by Klemperer and Meyer (1989) and Green and Newbery (1992). Further models in this vein include Bolle (1992), Bolle (2001) and Hobbs (2001). As emphasized by Bolle (2001), supply curve bidding will only lead to results different from traditional Cournot-Nash equilibria, if demand uncertainty (or another source of uncertainty) leads to an ex-ante undetermined equilibrium. Otherwise the supply (and demand curve) bidding collapses to one point, which corresponds to the Cournot-Nash equilibrium. Recent work in this field has on the one hand looked in more detail at the role of capacity constraints for strategic price equilibria (cf. e.g. von der Fehr, Harbord 1997; Baldick, Hogan 2002; Crampes, Creti 2002). On the other hand, numerous models have been set up to analyze the interactions between grid restrictions and market power for suppliers (e.g. Harvey, Hogan 2000, Hobbs et al. 2000, Gilbert et al. 2002, Metzler et al. 2003). Furthermore recent research has emphasized the role of a contract market for future and/or forward contracts for mitigating market power in electricity markets (cf. e.g. Newbery 1998, Ellersdorfer 2003).

Three points seem particularly worth retaining from this numerous literature: Most of the models developed are focusing on qualitative issues, taking quantitative results more as illustrative examples than as ultimate research objective. This is linked to the second observation that most analyses aim at providing decision support more to the regulators than to the utilities. The third point is that in most analyses, the entire focus is on the wholesale market and that the interlinkage between wholesale and retail markets is so far hardly analyzed.

2. COMPETITION ON THE RETAIL MARKET: BERTRAND COMPETITION AND COMPETITION WITH HETEROGENEOUS PRODUCTS

The competition on the retail market is quite different from the wholesale competition in that the firms are usually setting their sales prices and not the quantities sold. Notwithstanding the results (e.g. by Kreps, Scheinkman 1983) stating the equivalence of price (Bertrand) and quantity (Cournot) competition under certain circumstances, the analysis of the price competition on the retail market seems a worthwhile task. Thereby one has to take as starting point that besides prices further less directly observable characteristics of the commodities must influence the consumers' choices (e. g. service quality, billing frequencies, personal idiosyncrasies). This leads to a model of oligopolistic competition with product differentiation as a more adequate approach for describing observable behavior on competitive electricity markets.

Particularly in recent years major advances have been made towards a formal treatment of a broad class of such problems in a game theoretic framework (e. g. Economides, 1989; Caplin, Nalebuff, 1991; Anderson, de Palma, 1992; Anderson et al., 1995). Important application fields has been questions of locational choice in spatial economics, where products are not differentiated by their quality but by the location of the sales point and corresponding transport costs. Most applications to date yet focus on identical players with similar production structures. In the electricity markets on the contrary players of various sizes and with varying production assets do exist for historical reasons[24]. This holds both for countries with traditionally centralized electricity supply companies like France and the UK and for countries with a traditional oligopoly of regionally delimited power companies as it was in Germany or Sweden.

In the following the general methodological approach to model oligopolistic price competition in markets with differentiated goods is described first. The approach is then extended to multiple retail market segments and implications for optimal retail strategies are derived. Finally an application of the approach to the German electricity market is discussed.

2.1 Methodological approach – basic model with one retail market

We consider a non-cooperative game with N profit-maximizing players i (suppliers). These provide a bundle of differentiated commodities, with one commodity y_i supplied per player. The total demand y for the bundle of commodities is fix, but demand will shift among commodities depending on the prices p_i offered by the different players.

Various choices are possible for the individual demand functions y_i. We assume in the following a logistic shape of the functions y_i:

---

[24] In fact, this is also true for most other real markets.

$$y_i = D_i(p_1, p_2, ..., p_N, Y) = \frac{e^{\alpha_{0,i} - \alpha_{1,i} p_i}}{\sum_{j=1}^{N} e^{\alpha_{0,j} - \alpha_{1,j} p_j}} Y \qquad (5\text{-}1)$$

This functional form provides attractive features like lower and upper bounds of 0 and y respectively and a rather easy analytical treatment. Furthermore, it can be derived from a model of stochastic utility maximization (cf. Luce, 1959). Therefore, this type of model, called multinomial logit model, is widely used in the modeling of discrete choices (cf. e. g. Maddala, 1992)[25].

For notational convenience we introduce the market share m_i:

$$m_i = \frac{y_i}{Y} = \frac{e^{\alpha_{0,i} - \alpha_{1,i} p_i}}{\sum_{j=1}^{N} e^{\alpha_{0,j} - \alpha_{1,j} p_j}} \qquad (5\text{-}2)$$

Thereby $\alpha_{1,i}$ is a measure for the price responsiveness of the demand for the product offered by player i. $\alpha_{1,i}$ has to be positive to obtain normal demand functions. The larger the value of $\alpha_{1,i}$, the stronger is the price responsiveness.

$\alpha_{0,i}$ is a measure for the relative attractiveness of the product offered by player i. The larger $\alpha_{0,i}$, the higher the market share of commodity i at similar prices for all commodities. In fact, if $\alpha_{1,i} = \alpha_{1,j} = \alpha_1$ and $p_i = p_j = p$ (or more generally if $\alpha_{1,i} p_i = \alpha_{1,j} p_j$) for all i and j, one has:

$$m_i = \frac{e^{\alpha_{0,i}}}{\sum_{j=1}^{N} e^{\alpha_{0,j}}} \qquad (5\text{-}3)$$

In the non-cooperative game, each player i maximizes his profit G_i:

$$G_i = \left(p_i - \bar{c}_i(y_i)\right) y_i \qquad (5\text{-}4)$$

---

[25] Applications in economics and econometrics have been particularly pioneered by McFadden (1974, 1978). See also the application to heating equipment by Dubin and McFadden (1984).

by setting his sales price p_i, accounting for his average cost $\bar{c}_i(y_i)$, which correspond to marginal costs $c_i(y_i)$. Derivation accounting for Eq. (5-1) yields then:

$$\frac{\partial G_i}{\partial p_i} = \left(1 - \frac{c_i(y_i) - \bar{c}_i(y_i)}{y_i} \frac{\partial y_i}{\partial p_i}\right) y_i + (p_i - \bar{c}_i(y_i)) \frac{\partial y_i}{\partial p_i} \qquad (5\text{-}5)$$

Using Eqs. (5-1) and (5-3), the following identity is obtained after some transformations:

$$\frac{\partial y_i}{\partial p_i} = -\alpha_{1,i} y_i + \alpha_{1,i} \frac{y_i^2}{y} = -\alpha_{1,i} y_i (1 - m_i) \qquad (5\text{-}6)$$

and for the first order-conditions for a Nash-equilibrium:

$$\frac{\partial G_i}{\partial p_i} = 0$$

$$\Leftrightarrow y_i + (p_i - c_i(y_i)) \frac{\partial y_i}{\partial p_i} = 0 \qquad (5\text{-}7)$$

$$\Leftrightarrow y_i \left[1 - (p_i - c_i(y_i)) \alpha_{1,i} (1 - m_i)\right] = 0$$

If the possibility $y_i = 0$ (market exit) is excluded, prices and market shares in the competitive equilibrium must fulfill the identity:

$$\alpha_{1,i}(p_i - c_i(y_i))(1 - m_i) = 1 \qquad (5\text{-}8)$$

For constant marginal costs $c_i(y_i) \equiv c_i$, this is the equation of a hyperbole in the m_i- p_i-plane with asymptotes $p_i = c_i$ and $m_i = 1$ (see Figure 5-1). It corresponds to the optimal price setting of player i at a given market share m_i. The larger the absolute value of $\alpha_{1,i}$ (the price responsiveness of the customers), the closer the hyperbole remains to its asymptotes, i.e. the supply price exceeds the marginal cost only by little.

The demand function (5-2) can also be represented in this plane as a sigmoid with asymptotes $m_i = 0$ and $m_i = 1$ and the exact shape of the demand function depending on the parameter values and the prices of the other players. The demand is monotonously decreasing in p_i, but even for a price of 0 the player i will not have a market share of 1, given the incomplete substitutability of the different goods. The intersection of the hyperbolic pricing function $p_i(m_i)$ and the sigmoid demand function $D_i(m_i)$ yields a unique price-market-share combination for commodity i at any values of p_j ($j \neq i$). Since this holds for any commodity i and prices are always a strictly monotonous function $p_i(m_i)$ of the corresponding market shares, we obtain a unique solution for the first-order conditions (5-7). From this, the existence and uniqueness of a market equilibrium in this setting may be derived intuitively. A formal proof of the existence and uniqueness is given in Weber (2000)[26].

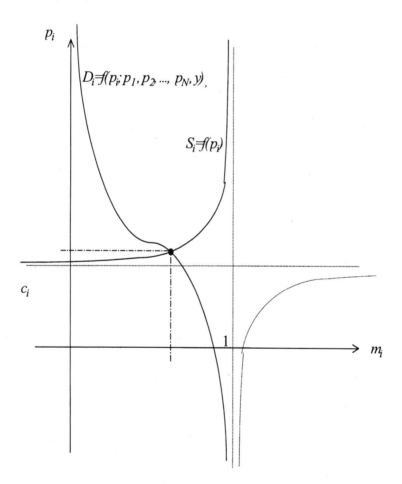

Figure 5-1. Demand function, optimal supply function and Nash-equilibrium in a non-cooperative game of oligopolistic competition

[26] For the symmetric case with constant marginal costs a proof may be found in Anderson / de Palma (1992).

Furthermore, it can be shown that the second derivative of the profit function G_i with respect to the own price p_i is negative when the first-order conditions are fulfilled. Hence profits are quasi-concave in own price and the obtained solution represents indeed profit maximums for all firms involved.

2.2 Methodological approach – extension to several retail segments

As before, a non-cooperative game with n profit-maximizing players is considered. But now n_Q market segments Q are taken into account, on which each player sells a differentiated product $y_{i,q}$. As before, the total demand y_q for each market segment is fixed, but depending on the prices $p_{i,q}$ offered, the players will obtain various market shares. The demand functions are now applied to each market segment, giving rise to the following equation for market shares $m_{i,q}$ within one segment:

$$m_{i,q} = \frac{y_{i,q}}{Y_q} = D_{i,q}(p_{1,q}, p_{2,q}, ..., p_{n,q}) = \frac{e^{\alpha_{0,i,q} - \alpha_q p_{i,q}}}{\sum\limits_{j=1}^{n} e^{\alpha_{0,j,q} - \alpha_q p_{j,q}}} \quad (5\text{-}9)$$

α_q is thereby again a measure of the price sensitivity in market segment q, but it is assumed to be equal for all players, whereas $\alpha_{0,i,q}$ indicates the relative attractiveness of product i in market segment s at equal prices. The profit of player i is now described by the equation:

$$G_i = \sum\limits_{q} p_{i,q} y_{i,q} - \int\limits_{0}^{\sum\limits_{q} y_{i,q}} c_i(\tilde{y}_i) d\tilde{y}_i \quad (5\text{-}10)$$

The decision variables of each player are now the sales prices $p_{i,q}$ in the different market segments. The costs are only a function of the current total sales quantity $y_{tot,i} = \sum\limits_{s} y_{i,s}$, i.e. retail and distribution costs are disregarded.

The first-order conditions for an equilibrium are now:

$$\frac{\partial G_i}{\partial p_{i,q}} = 0$$

$$\Leftrightarrow y_{i,q} + \left(p_{i,q} - c_i \left(\sum_q y_{i,q} \right) \right) \frac{\partial y_{i,q}}{\partial p_i} = 0 \tag{5-11}$$

$$\Leftrightarrow y_{i,q} \left[1 - \left(p_i - c_i \left(\sum_q y_{i,q} \right) \right) \alpha_q \left(1 - m_{i,q} \right) \right] = 0$$

Thereby the identity (5-6), generalized to the case of multiple market segments has been used again. Neglecting the possibility of market exit, this yields an optimality condition for the price in market segment q, which is very similar to Eq. (5-8):

$$p_{i,q} = c_i \left(\sum_q y_{i,q} \right) + \frac{1}{\alpha_q \left(1 - m_{i,q} \right)} \tag{5-12}$$

i.e. in each market segment, a supplier will charge a price which corresponds to his marginal generation costs plus a mark-up, which is independent of his generation costs, but rather depends on the price sensitivity in this market segment and the market share detained by the supplier in this segment.

The mark-up will be always higher if the market share is higher. But even if the market share, e.g. in a polypolistic competition, gets close to zero, the supplier will nevertheless charge a mark-up, proportional to the inverse of the price sensitivity parameter. In the opposite monopolistic case with $m_{i,q} = 1$ the optimal strategy for the supplier is to set an infinite price – as it is in general in a monopoly with zero (total) demand elasticity. But also in the competition case, the incumbents with high market shares will, at same cost structures, charge higher prices than newcomers with low market shares. This has also been observed empirically, both on the European markets for telecommunication services and for electricity. The still higher prices of the former monopolists in the telecommunication sector have according to this model to be viewed indeed as a profit optimizing strategy. In this market, the regulators in Germany and many other countries have obliged the grid operators to provide grid capacities at similar costs to all market participants. In the electricity market, besides the grid costs – which have also to be non-discriminatory – substantial costs arise also in the generation resp. wholesale purchase of electricity. Here only a numerical simulation can tell whether higher prices of the former regional monopolists are a consequence of cost differences or a result of market power.

At identical market shares, Eq. (5-12) implies that the mark-up will be the higher in a market segment, the lower the price sensitivity α_q of the demand side. Several analyses (e.g. VDEW 2000b) have shown that price differentials have to be considerably higher in the case of households and other small electricity customers to make them switch suppliers. This has according to Eq. (5-12) as immediate consequence that the mark-up beyond generation and transmission costs $c_i(q_i)$ will be in the competitive equilibrium higher for households than for more price sensitive customers.

But is it under these conditions attractive for a new competitor to enter the high-margin segment of another utility, or is it on the contrary non-rewarding, given the low switching willingness of the households? To answer this question, prices and quantities in the competitive equilibrium are scrutinized in more detail in the next section. Thereby the focus is on the case with two suppliers in two market segments, in order to be able to derive analytical results. Numerical results for the more general case of the German power market are presented in the subsequent section.

2.3 Analytical results – attractiveness of various market segments

For the case of two utilities competing in two market segments, the general equilibrium condition (5-12) may be written explicitly:

$$p_{1,A} = c_1\left(y_{1,A} + y_{1,B}\right) + \frac{Y_A}{\alpha_A\left(Y_A - y_{1,A}\right)}$$

$$p_{1,B} = c_1\left(y_{1,A} + y_{1,B}\right) + \frac{Y_B}{\alpha_B\left(Y_B - y_{1,B}\right)}$$

$$p_{2,A} = c_2\left(y_{2,A} + y_{2,B}\right) + \frac{Y_A}{\alpha_A\left(Y_A - y_{2,A}\right)} \qquad (5\text{-}13)$$

$$p_{2,B} = c_2\left(y_{2,A} + y_{2,B}\right) + \frac{Y_B}{\alpha_B\left(Y_B - y_{2,B}\right)}$$

Thereby the suppliers have been given the indices 1 and 2 and the market segments the indices A and B. Furthermore the equations have been formulated in quantities sold $y_{i,q}$ instead of market shares.

By taking differences we get:

$$p_{1,A} - p_{2,A} + \frac{Y_A}{\alpha_A(Y_A - y_{2,A})} - \frac{Y_A}{\alpha_A(Y_A - y_{1,A})} = c_1(y_{1,A} + y_{1,B}) - c_2(y_{2,A} + y_{2,B})$$

$$p_{1,B} - p_{2,B} + \frac{Y_B}{\alpha_B(Y_B - y_{2,B})} - \frac{Y_B}{\alpha_B(Y_B - y_{1,B})} = c_1(y_{1,A} + y_{1,B}) - c_2(y_{2,A} + y_{2,B})$$

$$(5\text{-}14)$$

From Eq. (5-9), the ratio between the quantities $y_{1,A}$ and $y_{2,A}$ can be determined. After taking logarithms, the following relationship is obtained:

$$\ln\left(\frac{y_{1,A}}{y_{2,A}}\right) = \ln\left(\frac{m_{0_{1,A}}}{m_{0_{2,A}}}\right) + \alpha_A(p_{2,A} - p_{1,A}) \tag{5-15}$$

Isolating the price difference and inserting it into Eq. (5-14) yields:

$$B_A(y_{1,A}) + c_1(y_{1,A} + y_{1,B}) = c_2(Y_A + Y_B - y_{1,A} - y_{1,B})$$

$$\text{with:} \quad B_A(y_{1,A}) = \frac{1}{\alpha_A}\left\{Y_A\left(\frac{1}{(Y_A - y_{1,A})} - \frac{1}{y_{1,A}}\right) + \ln\left(\frac{y_{1,A}}{Y_A - y_{1,A}}\right) - \ln\left(\frac{m_{0_{1,A}}}{1 - m_{0_{1,A}}}\right)\right\}$$

$$(5\text{-}16)$$

Consequently the marginal costs of the suppliers are not identical in the optimum – in contrast to the result established for perfect competition (cf. Chapter 4). Instead the marginal costs differ by the function $B_A(y_{1,A})$. This function may be interpreted as the shadow price of quantity increase, since similarly to Eq. (5-16) the following identities may be deduced:

$$B_B(y_{1,B}) + c_1(y_{1,A} + y_{1,B}) = c_2(Y_A + Y_B - y_{1,A} - y_{1,B})$$
$$B_A'(y_{2,A}) + c_2(y_{2,A} + y_{2,B}) = c_1(Y_A + Y_B - y_{2,A} - y_{2,B})$$
$$B_B'(y_{2,B}) + c_2(y_{2,A} + y_{2,B}) = c_1(Y_A + Y_B - y_{2,A} - y_{2,B})$$

$$\text{with:} \quad B_B(y_{1,B}) = \frac{1}{\alpha_B}\left\{Y_B\left(\frac{1}{(Y_B - y_{1,B})} - \frac{1}{y_{1,B}}\right) + \ln\left(\frac{Y_B - y_{1,B}}{y_{1,B}}\right) - \ln\left(\frac{1 - m_{01,B}}{m_{01,B}}\right)\right\}$$

$$B_A'(y_{2,A}) = -B_A(y_{1,A})$$
$$B_B'(y_{2,B}) = -B_B(y_{1,B})$$

$$(5\text{-}17)$$

By comparing Eq. (5-16) and the first equation in (5-17), the equality $B_A(y_{1,A}) = B_B(y_{1,B})$ is readily established as a further characteristic of market equilibrium. The decision in which market segment firm 1 (or 2) should chase for new customers, has therefore to be taken in a way, that the shadow price of customer acquisition (or quantity extension) is the same in both market segments. This shadow price is not depending on the cost structure of the firm or of its competitor, it is on the contrary a function of the market characteristics, namely of price sensitivity α_q, market volume Y_q and market shares at identical price $m_{0,i,q}$.

From Eqs. (5-16) and (5-17), it can be shown that the shadow prices $B_A(y_{1,A})$ und $B_B(y_{1,B})$ are strictly increasing functions of the quantities sold, i.e. with increasing market share, it gets increasingly difficult to get additional customers. A selection of shadow price curves for selected parameter values is shown in Figure 5-2.

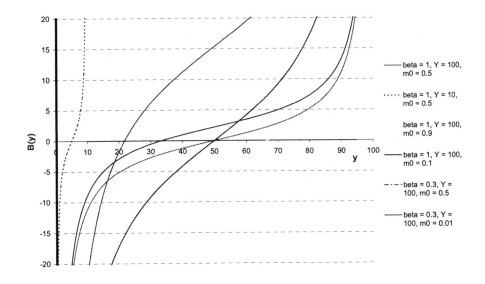

Figure 5-2. Shadow price of quantity increases depending on market characteristics

The customer acquisition gets clearly more difficult, the lower the price sensitivity is. By contrast, differences in the market share at equal prices have less drastic impacts on the shadow price for quantity increases.

Figure 5-3 shows how the shadow price curves for different market segments may be aggregated horizontally. This is valid, since the quantity extensions possible at a given shadow price level are obtained as the sum of the quantity extensions in the different market segments. Subsequently a (vertical) aggregation with the cost function is also possible, so that the optimal competition result may be determined graphically.

3.   APPLICATION

An empirical application of the model is done for a simplified version of the German power market. In the following, the key parameters for this model are described first and then numerical simulation results are provided.

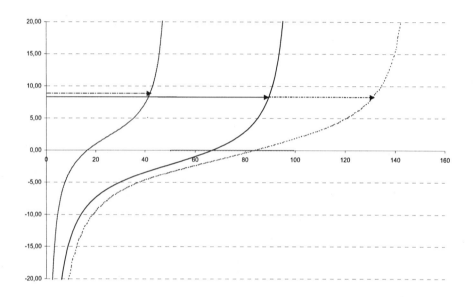

Figure 5-3. Aggregation of the shadow prices functions for quantity extensions

3.1   Market description

Instead of modeling the about 1000 utilities in the German market, the focus is laid on the largest suppliers, i.e. the so-called "Verbundunter-nehmen" (national utilities) and those regional and municipal utilities which supply more than two third of their sales to final customers through own generation. This gives a total of 26 players according to the situation before the mergers of the years 2000 to 2002, and 22 thereafter. The generation capacities are modeled similarly to Ellersdorfer et al. (2000) and the model is formulated in annual energy quantities (TeraWatthours, TWh). Given that the electricity demand of the consumers is not equally spread over the year, the energy supply requires the combination of different production technologies. Consequently the marginal generation cost for one additional TWh is also not given by one production technology, but by a generation mix which includes the marginal production technologies at different moments within one year. Ellersdorfer et al. (2000) have developed an approach to tackle this difficulty based on the assumption of constant load shape and load factor. This approach is also used here. Thereby the production technologies are distinguished by primary fuel used and age category and marginal production cost curves as shown in Figure 5-4 are derived through a Gompertz function approximation. For all utilities, not only their own generation capacities are included, but the generation capacities of companies fully or partially under the control of the companies considered.

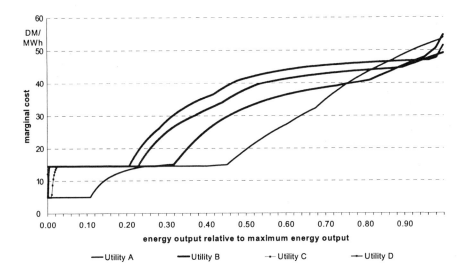

Figure 5-4. Marginal generation cost for energy output for selected electricity suppliers in Germany

On the demand side, three market segments are considered: the retailers, which mostly supply electricity to final customers without having substantial own generation capacities, the larger industrial and commercial customers and the residential and small business clients. The quantities sold to these customers are taken from VDEW (2000a). At the same time, these quantities are used to calibrate the model parameter "market share at equal prices" (cf. Figure 5-5). Thereby on the one hand the market share before liberalization is used, considering that customers have little incentives to switch suppliers at equal prices. Additionally it is assumed that the large companies are better known (or can make themselves better known) and that consequently clients, who switch suppliers for non-price reasons (such as service quality, billing modalities etc.), generally will rather switch from the small to the larger companies than vice versa. The parameter "market share at equal prices" accounts for these two factors through a weighted average. 95 % of the weight is accorded to the original market share in the market segment considered and 5 % to the total electricity supply as indicator of the size of the company[27]. The market shares given are always with respect to those customers directly supplied by the companies under consideration.

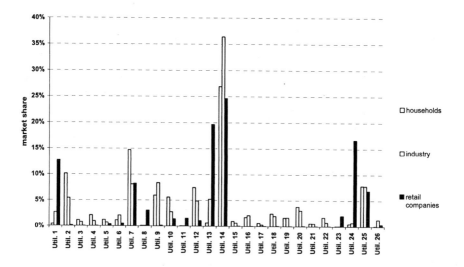

Figure 5-5. Market share at identical prices of the electric utilities considered

For the price responsiveness parameter α_q, the values have been chosen on the basis of results of various surveys (Stern 1999, Stern 2000, VDEW 2000b). Thereby the highest value is attributed to retail companies and the lowest to private households.

Furthermore transmission and distribution costs have also to be taken into account when modeling competition on the retail side. Two cases are considered here:

- For *residential customers*, average transport and distribution costs (including eco-tax and VAT) of 120 €/MWh and
- for *industrial and commercial customers*, transport and distribution costs of 24 €/MWh

---

[27] Obviously an empirical investigation determining the appropriate parameter values would be very desirable. However, this is hardly feasible on the basis of publicly available data in Germany.

are taken. Transport and distribution costs are assumed to be constant and similar for all companies, given that transport fees have to be non-discriminative and that for distribution costs no detailed information is available.

3.2   Results

Figure 5-6 shows the changes in market shares for the different suppliers in the Nash-equilibrium compared to the market share at equal prices. Clearly for most producers the changes in market shares are rather limited. Yet the largest producer exhibits a decrease in the sales quantity by approximately three percentage points in two out of the three market segments considered. This is a consequence of the market power of this producer, which allows him to charge higher prices than the other producer. But already for the second largest producer, the demand increases in two out of the three market segments. This is particularly a consequence of the comparatively low marginal costs of this supplier (cf. Figure 5-9). Also among the other suppliers, those show mostly sales increases which have a competitive costs structure. The correlation coefficient between the total change in electricity sold (cf. Figure 5-7) and the marginal costs (cf. Figure 5-9) is as large as −0.84.

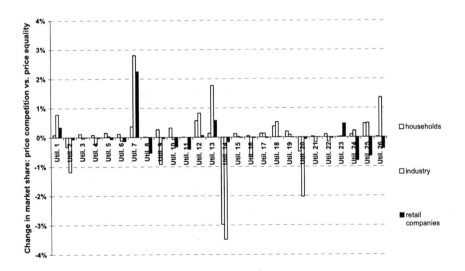

Figure 5-6. Changes in market shares under price competition compared to the situation under equal prices

The example of the second largest supplier also illustrates that it is despite higher mark-ups not necessarily attractive to enter the market for household customers. Given the low price sensitivity, high shadow prices have to be paid (or price concessions to be made, cf. Figure 5-2) to make deals with the private customers. The incumbents may on the other hand reduce substantially the business opportunities for the new player if they reduce in turn their prices – a behavior which has clearly been observed in Germany in the first years after liberalization.

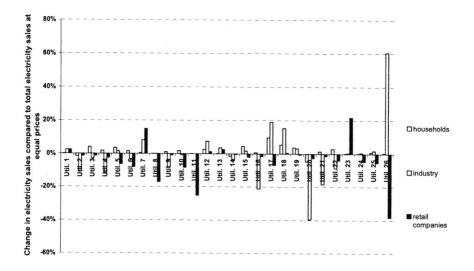

Figure 5-7. Changes in electricity production under price competition

Figure 5-8 illustrates that price differentials are highest in the equilibrium within the market segment of residential customers, given the low price sensitivity in this field. Conversely the high price sensitivity of distribution companies and industry induces very similar offers by the utilities and their mark-ups are substantially reduced. This is similar to the Ramsey-pricing rule for combined production, according to which the costs of joint production should be distributed according to the price elasticity of the demands on the different products.

Compared with the initial situation before liberalization, especially the utilities 11, 16 and 20 find their demand considerably reduced (Figure 5-7). These are throughout smaller suppliers with comparatively high production costs. In general the correlation between the size of the utility and the change in the electricity sold is however low, averaged over all utilities it is only 0.08. Much higher is on the contrary, as mentioned earlier, the correlation between the changes in electricity supplied and the marginal costs of

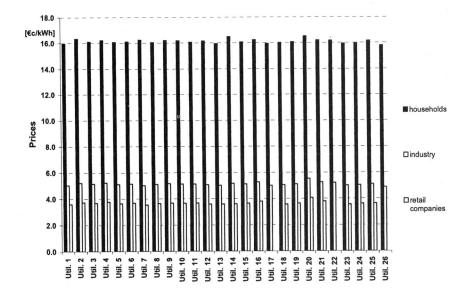

Figure 5-8. Prices under price competition in the different market segments

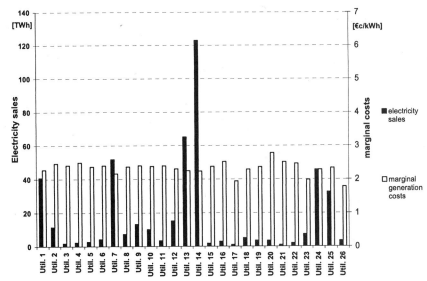

Figure 5-9. Electricity sales and marginal costs of generation in price competition

generation[28]: The suppliers with the highest growth in output, notably

---

[28] The costs for transmission and retail services have been assumed to be equal for all players. This is true for transmission and distribution grid fees due to the discrimination

utilities 17, 23 and 26, have at the same time also the lowest marginal costs, providing them for example a mark-up of 6.5 €/MWh to 7.2 €/MWh in their sales to industrial customers, which is much higher than for the average of all suppliers (4.3 €/MWh).

Even in their trades with retailing utilities, their margin is comparatively high. But also for the largest supplier, the mark-up beyond marginal costs is with 5.5 €/MWh for industrial customers higher than average. This indicates that both market power resp. customer loyalty and cost-effective production contribute to the commercial success of vertically integrated electric utilities in the liberalized markets. Through the model presented here, it is possible to formulate this statement not only verbally and abstractly, but also to make quantitative analyses and to identify the relative importance of these different success factors.

On the whole, the presented model of oligopolistic price competition offers a specific perspective on the retail competition of vertically integrated companies. Although the numeric results obtained should not be overinterpreted, they clearly show the important role both of competitive cost structures and market power respectively customer loyalty for obtaining positive commercial results. It turns also out that substantial extra profits for the suppliers are still possible even in a competitive environment, especially in customer segments with low price sensitivity. It has however been shown both theoretically and numerically that this market segment is not necessarily an attractive field for expansion plans, since customer switching can only be obtained at high costs and competitors may well counteract acquisition efforts through own price reductions.

As with many game theoretic models, the essential results are more qualitative than quantitative and clearly further investigations would be needed, notably to obtain reliable estimates on the price sensitivity of the customers. Then the tool could also become a worthy tool for quantitative analysis. On the other hand, the derived global regularity properties make the basic model also an attractive candidate for further refinement and extension, e.g. for the inclusion of pure retail companies as separate players and for an analysis of the interaction between wholesales and retail markets.

---

free grid access guaranteed by German and European law, yet for the retail business this is an approximation, which could be improved if better data were available.

Chapter 6

OPTIMIZING GENERATION AND TRADING PORTFOLIOS

Unit commitment and load dispatch have been key issues in the electricity industry ever since the establishment of large integrated networks (cf. e.g. the literature surveys by Sheble and Fahd 1994, Sen and Kothari 1998, Römisch 2001, Wallace and Fleten 2002 and the recent volumes by Hobbs et al. 2001 and Bhattacharya et al. 2001). Yet, the problem has been considerably modified through liberalization and unit commitment has to be rethought under the new conditions. In particular, price uncertainty has now to be accounted explicitly for (cf. e.g. Krasenbrink et al. 1999, Takriti et al. 2000, Dotzauer, Lindberg 2001, Brand, Weber 2001). This has provided an additional push towards the use of stochastic optimization approaches (cf. also Dentcheva and Römisch 1998, Gröwe-Kuska et al. 1999, Conejo et al. 1999, Bacaud et al. 2001 and others). But so far only limited attention has been devoted to develop models which fit precisely to the specific market context where they are to be used.

Taking the market structure described in chapter 2 as starting point, the following decision problems have notably to be solved by generators:

- longer-term portfolio management, balancing derivative market products and physical optionalities given by the power plants on the spot market,
- bid generation for the physical day ahead market,
- bid generation for the reserve power market,
- unit commitment after the closure of the day ahead and the reserve power market.

In the following, the focus will be laid on the first and the last of these problems, because these are the most important ones in terms of financial flows and physical commitments. Furthermore, the bid generation for the spot market is dealt with extensively by Brand et al. (2002), Brand, Weber

(2002) and Brand (2004), whereas the optimal bid calculation for reserve power markets is discussed by Swider, Weber (2003b) and Swider (2004).

Besides the emphasis on models adapted to the specific market structure, a particular focus is laid in the following on CHP systems, not the last given the current strive for distributed generation and efficient energy use. Consequently, the technical elements which have to be accounted for in the production scheduling of CHP plants are first described in section 1. Thereby a formulation suitable for planning under uncertainty is chosen from the outset. The system-wide restrictions and the choice of the objective function – depending on the market context - are then discussed in section 2. Next, specific types of decision problems are dealt with. Section 3 is devoted to separable optimization problems, as they arise in liberalized markets for non-CHP systems. Section 4 describes an approach for two-stage optimization of CHP systems in the day-ahead market. Section 5 looks then at the longer-term portfolio management problem for these systems, whereas the application of these methods is discussed in section 6.

1.   TECHNICAL ELEMENTS FOR PRODUCTION SCHEDULING

Key elements for describing the operation of all types of thermal power plants are fuel consumption equations and capacity constraints (section 1.1), together with start-up and shut down constraints (cf. section 1.2). Specific elements for CHP plants depend on the type of the units considered. Here backpressure steam turbines (cf. section 1.3), extraction-condensing plants (cf. section 1.4) and gas turbines (section 1.5) are considered. The description is based on earlier work by Hanselmann (1996), Albiger (1998) and Bagemihl (2001) and follows in general the treatment in Thorin et al. (2001). Similar descriptions may also be found e.g. in Hönig (2001) or Dotzauer (2001). But here in view of the definition of a stochastic optimization problem, not only indexation over time is done, but also on scenarios. A scenario S consists thereby of a sequence of states of the world s_t for the different time steps t, formally:

$$S = \left\{ s_t \middle| t \in [t_0\ t_{max}] \wedge \forall (s, s') \in \mathbf{S}_t \times \mathbf{S}_{t+1} \Pr(s \rightarrow s') > 0 \right\} \qquad (6\text{-}1)$$

In order to streamline the notation somewhat, the index t which should accompany the state s_t is suppressed in the following. Non-anticipativity as defined e.g. in Römisch (2001) is thereby assumed for decisions in the different scenarios. I.e. if a state s_t belongs to different scenarios S (since those diverge only later in time), then any decision at s_t has to be identical for all scenarios S.

Throughout, equations are expressed in energy terms, so no specific attention is paid to mass flows and temperature levels. But a wide range of CHP plants can be adequately described through this approach. The major exception are CHP systems with several steam extraction and collection levels, where mass balances and temperature levels allow to capture more adequately the thermodynamic aspects and the resulting restrictions for plant operation (Hanselmann 1996). The level of modeling detail is far lower than in single-plant simulation programs (e.g. Schnell u. a. 2002), but here a level of aggregation is chosen, which allows to solve afterwards the overall unit commitment problem in limited computation time. This is also the reason, why non-linear equations are replaced by their (piecewise) linearised counterparts.

1.1 Fuel consumption and capacity restrictions for conventional power plants and boilers

The fundamental equations to describe a thermal power plant without combined heat production are firstly restrictions on maximum and minimum electric power as given by the power plant characteristics.

Maximum electric power:

$$P_{u,t,s}^{EL} \le P_{u,\max}^{EL} O_{u,t,s} \qquad (6\text{-}2)$$

Minimum electric power:

$$P_{u,t,s}^{EL} \ge P_{u,\min}^{EL} O_{u,t,s} \qquad (6\text{-}3)$$

Thereby $O_{u,t,s}$ describes the on-off status of the plant. Thus the upper bound for electric power is zero if the plant is turned off and $P_{EL,u}^{\max}$ if the plant is turned on. Similarly, the lower bound for output power is only different from zero if the plant is turned on. These equations have obviously to be fulfilled for all units u, all time steps t and all states s at these time steps. These precisions are omitted in the above equations and will also be omitted in the following to improve readability, except when of primordial importance.

Secondly, the fuel consumption equation describes the relationship between fuel input $\dot{m}_{f,u,t,s}$ and electricity output $P_{u,t,s}^{EL}$:

$$\dot{m}^{FU}_{f,u,t,s} = \frac{P^{EL}_{u,t,s}}{\tilde{h}_f \eta_u (P^{EL}_{u,t,s})} \tag{6-4}$$

where $\tilde{h}_f$ is the fuel calorific value and η_u is the load dependent efficiency of the plant.

This equation can be linearised and directly formulated using the fuel input power $P^{FU}_{f,u,t,s}$ as variable to obtain a more tractable problem:

$$P^{FU}_{f,u,t,s} = \beta^{EL}_u P^{EL}_{u,t,s} + \beta^O_u O_{u,t,s} \tag{6-5}$$

It should be noted that this is an affine equation where the second term is non-zero if the plant is operating. Thus the linearisation is done in a way to obtain an optimal fit only on the interval $\left[P^{EL}_{u,min} P^{EL}_{u,max}\right]$ and the coefficient β^{EL}_u represents the marginal fuel consumption when the plant is operating, whereas β^O_u is the fuel consumption at minimum output which exceeds the quantity $\beta^{EL}_u P^{EL}_{u,min}$.

For heat producing boilers, which are often included as peak load devices in CHP systems, the relation between fuel consumption $\dot{m}^{FU}_{f,u,t,s}$ and boiler heat output flow $\dot{Q}^{HT}_{u,t,s}$

$$\dot{m}^{FU}_{f,u,t,s} = \frac{\dot{Q}^{HT}_{u,t,s}}{\tilde{h}_f \eta_u (\dot{Q}^{HT}_{u,t,s})} \tag{6-6}$$

can be expressed similarly as:

$$P^{FU}_{f,u,t,s} = \beta^{TH}_u \dot{Q}^{HT}_{u,t,s} + \beta^O_u O_{u,t,s} \tag{6-7}$$

The boiler output is also limited by the technical minimum and maximum power:

$$\dot{Q}^{HT}_{u,min} O_{u,t,s} \leq \dot{Q}^{HT}_{u,t,s} \leq \dot{Q}^{HT}_{u,max} O_{u,t,s} \tag{6-8}$$

For the mathematical structure of the problem, boilers and power plants are therefore equivalent, except that they provide different outputs.

1.2   Start-up, shut-down and ramping constraints

Besides the restrictions for fuel consumption and maximum and minimum power, also further technical restrictions for electric and thermal power are to be satisfied by all plant types at each time point. These include notably minimum operation times $T_{u,\min}^{UP}$ and minimum down times $T_{u,\min}^{DOWN}$, but also maximum ramping rates.

For a compact notation of these restrictions (and also later on for the modeling of start-up and shut-down costs) binary variables $U_{u,t,s}$ and $D_{u,t,s}$ describing respectively the start-up and the shut-down of a unit are introduced. These are binary variables which are defined through the following restrictions: $U_{u,t,s}$ must have the value 1, whenever the state of the unit changes from offline to online, i.e. the variable $O_{u,t,s}$ switches from 0 to 1 within one scenario S:

$$U_{u,t,s} \geq O_{u,t,s} - O_{u,t-1,s'} \qquad \forall S \; \forall s, s' \in S \qquad (6\text{-}9)$$

In all other cases, the start-up variable has to be zero; therefore it is bounded from above by the current operation state and also bounded from above by the opposite of the operation state at the previous time step:

$$U_{u,t,s} \leq O_{u,t,s} \qquad \forall S \; \forall s \in S \qquad (6\text{-}10)$$

$$U_{u,t,s} \leq 1 - O_{u,t-1,s'} \qquad \forall S \; \forall s, s' \in S \qquad (6\text{-}11)$$

Similar considerations are valid for the shut-down process and lead to the restrictions:

$$D_{u,t,s} \geq O_{u,t-1,s'} - O_{u,t,s} \qquad \forall S \; \forall s, s' \in S \qquad (6\text{-}12)$$

$$D_{u,t,s} \leq O_{u,t-1,s'} \qquad \forall S \; \forall s, s' \in S \qquad (6\text{-}13)$$

$$D_{u,t,s} \leq 1 - O_{u,t,s} \qquad \forall S \; \forall s \in S \qquad (6\text{-}14)$$

Restrictions on start-up or shut-down frequency are mostly introduced to prevent increased risks of component failures and are particularly relevant for coal-fired or nuclear plants. Therefore a minimum down time $T_{u,\min}^{DOWN}$ is defined for those units u. Likewise it is not possible to shut down a unit if it was started-up the time-step before, and a minimum operation time $T_{u,\min}^{UP}$ is defined. A unit can then only be shut-down after $T_{u,\min}^{UP}$ hours of operation, which can be written formally:

$$U_{u,t,s}T_{u,\min}^{UP} \leq \sum_{\tau=t}^{t+T_{u,\min}^{UP}-1} O_{u,\tau,s'} \qquad \forall S \; \forall s,s' \in S \qquad (6\text{-}15)$$

Similarly, one gets for minimum down times:

$$D_{u,t,s}T_{u,\min}^{DOWN} \leq \sum_{\tau=t}^{t+T_{u,\min}^{DOWN}-1} \left(1-O_{u,\tau,s'}\right) \qquad \forall S \; \forall s,s' \in S \qquad (6\text{-}16)$$

Additional restrictions are sometimes imposed in form of ramp rates, again to prevent excessive material and component fatigue. Thus power and heat gradients may not exceed certain values during operation. However, the situations at start up and shut down of each unit have to be accounted for since during these special events, the gradient is allowed to be higher than during the operation of the unit. This requires the introduction of the parameter $P_{EL,u}^{*} = \max(0, P_{EL,u}^{\min} - P_{EL,u}^{\max.Change})$, which is associated then with the start-up resp. the shut-down event. Thus for electric power one gets:

$$P_{EL,u,t,s} - P_{EL,u,t-1,s'} - U_{u,t,s}P_{EL,u}^{*} \leq P_{EL,u}^{\max.Change} \qquad \forall S \; \forall s,s' \in S \qquad (6\text{-}17)$$

$$P_{EL,u,t-1,s'} - P_{EL,u,t,s} - D_{u,t,s}P_{EL,u}^{*} \leq P_{EL,u}^{\max.Change} \qquad \forall S \; \forall s,s' \in S \qquad (6\text{-}18)$$

Similar equations may be formulated for heat output:

$$\dot{Q}_{u,t,s} - \dot{Q}_{u,t-1,s'} - U_{u,t,s}\dot{Q}_{u}^{*} \leq \dot{Q}_{u}^{\max.Change} \qquad \forall S \; \forall s,s' \in S \qquad (6\text{-}19)$$

$$\dot{Q}_{u,t-1,s'} - \dot{Q}_{u,t,s} - D_{u}\dot{Q}_{u}^{*} \leq \dot{Q}_{u}^{\max.Change} \qquad \forall S \; \forall s,s' \in S \qquad (6\text{-}20)$$

with $\quad \dot{Q}_{u}^{*} = \max\left(0, \dot{Q}_{u}^{\min} - \dot{Q}_{u}^{\max.Change}\right)$

1.3   Back-pressure steam turbines

Back-pressure steam turbines are often used in CHP if power and heat are needed simultaneously and in rather stable shares, since they produce electricity and heat in a constant ratio. In the following, a back-pressure turbine is modeled together with its corresponding boiler. Thereby the relations between output heat flow $\dot{Q}^{HT}_{u,t,s}$, output electric power $P^{EL}_{u,t,s}$ and input power $P^{FU}_{f,u,t,s}$ can be written in linearised form as follows:

$$P^{EL}_{u,t,s} = s^{EL}_u \dot{Q}^{HT}_{u,t,s} \tag{6-21}$$

$$P^{FU}_{f,u,t,s} = \beta^{EL}_u P^{El}_{u,t,s} + \beta^{O}_u O_{u,t,s} \tag{6-22}$$

s^{EL}_u thereby describes the power to heat ratio, which is a plant specific characteristic, β^{EL}_u is as before the marginal fuel consumption rate per unit of electricity output and β^{o}_u the fuel consumption at minimum output.

Additionally, power and heat output are again restricted by upper and lower bounds. Since they are directly linked through Eq. (6-21), it is sufficient to write restrictions on one output:

$$P^{El}_{u,\min} O_{u,t,s} \leq P^{El}_{u,t,s} \leq P^{El}_{u,\max} O_{u,t,s} \tag{6-23}$$

1.4   Extraction condensing steam turbine

Extraction-condensing steam turbines are a more complex turbine type, which is rather frequently used in combined heat and power generation due to its flexibility. Such turbines offer the possibility to produce power and heat in an at least partly variable ratio, by extracting steam (possibly at different stages) of a conventional (set of) steam turbine. For our purposes, we consider only the aggregated heat and power flows generated by the extraction-condensing turbine. The plant operation may then be described graphically as depicted in Figure 6-1 by a fuel consumption curve and a power and heat chart (brief: P-Q-chart), which indicates the operation limits for the combined heat and power production.

Formally the fuel consumption can be expressed through the equation:

$$P_{f,u,t,s}^{FU} = \frac{\dot{Q}_{u,t,s}^{STM}}{\eta_u\left(\dot{Q}_{u,t,s}^{STM}\right)} = \beta_u^{EL} P_{u,t,s}^{EL} + \beta_u^{HT} \dot{Q}_{u,t,s}^{HT} + \beta_u^{O} O_{u,t,s} \qquad (6\text{-}24)$$

The fuel consumption, which is prima facie depending on the steam output from the boiler (equal to the steam inlet to the turbine), is hence in a second step decomposed in a marginal fuel consumption β_u^{EL} for electricity production, a marginal fuel consumption β_u^{HT} for heat production and a fuel consumption β_u^{O} at minimum output. In Figure 6-1, the fuel consumption at

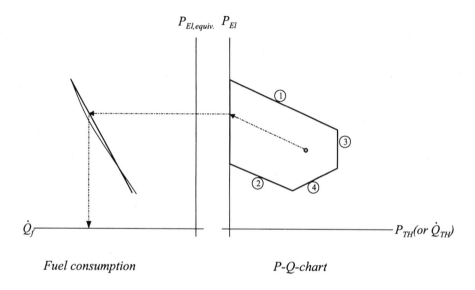

Fuel consumption   *P-Q-chart*

Figure 6-1. Condensing extraction turbine – graphical representation of fuel consumption and operation restrictions

the operation point marked by a dot is determined graphically from the equivalent electricity production $P_{El,equiv} = P_{El} + \beta_u^{HT} / \beta_u^{EL} \dot{Q}$, but the underlying relationship is clearly equivalent. The ratio $\beta_u^{HT} / \beta_u^{EL}$ thereby corresponds to the electric power reduction due to heat production.

The operation restrictions can be described by the inequalities:

$$\beta_u^{EL\,\prime} P_{u,t,s}^{EL} + \beta_u^{HT\,\prime} \dot{Q}_{u,t,s}^{HT} \leq \dot{Q}_{u,max}^{STM} \qquad (6\text{-}25)$$

$$\beta_u^{EL\,\prime\prime} P_{u,t,s}^{EL} + \beta_u^{HT\,\prime\prime} \dot{Q}_{u,t,s}^{HT} \geq \dot{Q}_{u,min}^{STM} \qquad (6\text{-}26)$$

$$\dot{Q}_{u,t,s}^{HT} \leq \dot{Q}_{u,\max}^{HT} \tag{6-27}$$

$$P_{u,t,s}^{EL} \geq s_{u,\min}^{EL} \dot{Q}_{u,t,s}^{HT} \tag{6-28}$$

The restriction (6-25) corresponds thereby to line①in Figure 6-1, which represents the maximum steam output from the boiler. Restriction (6-26) conversely describes the minimum output from the boiler, corresponding to line②in Figure 6-1. The coefficients $\beta_u^{EL\,\prime}, \beta_u^{EL\,\prime\prime}$ and $\beta_u^{HT\,\prime}, \beta_u^{HT\,\prime\prime}$ can thereby in many cases be set equal to their counterparts in the linearised fuel consumption equation β_u^{EL} and β_u^{HT}. The third operation restriction, inequality (6-27) resp. line③ is due to the limited heat exchanger capacity, which implies a maximum heat outlet. Finally, restriction (6-28) (line ④) provides a minimum power to heat ratio $s_{u,\min}^{EL}$, which is caused by the maximum ratio between the extraction and the condensing flow. One has to bear in mind that all these bounds, except maximal and minimal power generated, depend in principle on the temperature levels in the corresponding heating grid and may therefore vary between seasons.

1.5   Gas turbine and gas motor

Besides steam turbines, also gas turbines and gas motors are frequently used in CHP systems, notably in smaller ones. The model descriptions of the gas turbine and gas motor are similar to the model of the back-pressure steam turbine. The relation between the output heat flow $\dot{Q}_{u,t,s}^{HT}$ and the electric power output $P_{u,t,s}^{EL}$ can be expressed with a linear relation:

$$P_{u,t,s}^{EL} \geq s_u^{EL} \dot{Q}_{u,t,s}^{HT} \tag{6-29}$$

Here an inequality is used, describing the case where the heat generated by the motor or gas turbine can alternatively be taken through a heat exchanger to produce useful heat or is directly released to the environment through some auxiliary cooling system. If no auxiliary cooling system is present, the equality (6-21) should be used instead.

Fuel consumption is then again described through a (linearised) function of power production:

$$P_{f,u,t,s}^{FU} = \beta_u^{EL} P_{u,t,s}^{El} + \beta_u^{O} O_{u,t,s} \tag{6-30}$$

and upper and lower bounds for power production are formulated as before:

$$P_{u,\min}^{El} O_{u,t,s} \le P_{u,t,s}^{El} \le P_{u,\max}^{El} O_{u,t,s} \tag{6-31}$$

Separate restrictions for the heat output are in general not necessary, since it has an upper bound given through (6-29). Only in the case of an additional limitation through the heat exchanger capacity a further restriction is required. The equations for a gas motor differ from those for a gas turbine only by the parameter values and the fact that the parameters do not depend on the outdoor temperature.

2. SYSTEM-WIDE RESTRICTIONS AND OBJECTIVE FUNCTION

The operation of power plants and CHP plants is in general not only determined by the technical restrictions of each single plant but also through system wide load restrictions and the aim of the operators to maximize their profits. Before liberalization, sales revenues were mostly given in advance by the fixed rates and the (almost) fixed load, so unit commitment and load dispatch were usually solved as cost minimization problems with load constraints to be fulfilled. In the current situation, the balance of supply and demand is still an important aspect which is discussed in section 2.1. The objective function to be used for unit commitment and load dispatch decisions is then formulated in section 2.2. Section 2.3 then analyses the problem structure obtained and possible simplifications, establishing thus the link to the subsequent sections.

2.1 Balance of supply and demand

Prior to liberalization, the balance of supply and demand to be achieved for electric power mostly meant that the sum of the electricity produced by all units in the system had to equal the electric load. In liberalized markets, the production from own units may be complemented by purchases from the market – either through longer term contracts or at the spot market. On the other hand, electricity may also be sold through contracts and at the spot market. Thus one obtains the generalized inequality for the electrical load at each time t and in each scenario/state of the world s[29]:

$$\sum_{u \in EU} P_{u,t,s}^{EL} + \sum_{c \in BEC_t} P_{BUY,c,t,s}^{EL} \geq \sum_{c \in SEC_t} P_{SELL,c,t,s}^{EL} + P_{Spot,t,s}^{EL} \qquad (6\text{-}32)$$

Thereby, $P_{EL,u}(t)$ is the power produced by unit u, $P_{BUY,c,t,s}^{EL}$ describes the power bought under some contract c, $P_{SELL,c,t,s}^{EL}$ is the power sold at time t under contract c and $P_{Spot,t,s}^{EL}$ describes the quantity purchased at or sold to the spot market. Hence, $P_{Spot,t,s}^{EL}$ can be here, contrarily to the other variables, positive or negative.

Since a short-term market does generally not exist for heat, the balance equation is slightly different for heat: the heat output of all units in the system plus eventual heat purchases should at least equal the heat demand:

$$\sum_{u \in HU_i} \dot{Q}_{u,t,s}^{HT} + \sum_{c \in BTC_{g,t,s}} \dot{Q}_{BUY,c,t,s}^{HT} \geq \sum_{c \in STC_{g,t,s}} \dot{Q}_{SELL,c,t,s}^{HT} \qquad \forall g \in \mathbf{G} \qquad (6\text{-}33)$$

with:

$\dot{Q}_{u,t,s}^{HT}$ - heat produced by unit u at time t

$\dot{Q}_{BUY,c,t,s}^{HT}$ - heat bought under contract c for heat grid g at time t

$\dot{Q}_{SELL,c,t,s}^{HT}$ - heat sold under contract c in heat grid g at time t

In both inequalities, units that can produce electric or thermal power have to be counted according to their actual output. Units that produce both electric and thermal power thus appear in both inequalities.

The fuels used for the units are provided either by immediate delivery (e.g. from the gas grid) or taken from storage. In order to account for supply and demand contracts for fuels, the following balance equation is formulated:

$$M_{f,t,s} + \sum_{u \in U} P_{f,u,t,s}^{FU} + \sum_{c \in SFC_{g,t}} P_{SELL,f,c,t,s}^{FU} \leq M_{f,t-1,s'} + \sum_{c \in BFC_{g,t}} P_{BUY,f,c,t,s}^{FU} \qquad (6\text{-}34)$$

---

[29] Due to the technical restrictions mentioned above, the amounts of electricity produced can not be chosen arbitrarily small (e.g. minimum electric power production). Thus demanding strict equality would possibly lead to non-solvable problems.

Thereby fuel storage is accounted for through the storage quantities $M_{f,t-1,s}$, respectively $M_{f,t,s}$.

Besides these balances for electricity, heat and fuels, also balances for reserve power could be established, if the system under consideration has to provide these services. In that case, which is not analyzed in depth here, one would also have to adapt the capacity restrictions for the plants providing ancillary services and the provision of these services would have to be modeled (cf. chapter 4 and Thorin et al. 2003).

2.2   Objective function

In a deterministic setting, the aim of a power plant system operator clearly is to maximize the total profit of the system under consideration. With uncertainty (or more precisely risk), the operator may do some trade-off between increasing the expected value of the profit and limiting the risk. But in the following, we will limit ourselves to the risk-neutral case[30]. Hence, the objective function G can be defined as the difference between total expected revenues V and total expected costs C, with possibly some correction term to account for changes in stock ΔM.

$$\max G = \max \{V - C - \Delta M\} \tag{6-35}$$

The maximum is taken over all decision variables describing the operation of all units of the power plant system. Formulated this way, the objective function is basically valid both for conventional and market driven operation of power plant systems. And it is also applicable for deterministic and stochastic problem formulations.

Besides sales of electricity $V_{t,s}^{EL}$, heat sales $V_{t,s}^{HT}$ and possibly (re-)sales of fuels $V_{t,s}^{FU}$ contribute also to the total revenues V for CHP systems. Under risk neutrality, one has thus:

$$V = \sum_{t=1}^{T} \sum_{s \in S_t} \Pr_s \left[V_{t,s}^{EL} + V_{t,s}^{HT} + V_{t,s}^{FU} \right] \tag{6-36}$$

The value of the sales can be determined from the quantities sold, multiplied with the (possibly) time- and state-dependent prices:

---

[30] Methods for solving stochastic mixed integer linear programs with risk aversion are developed by Schultz, Tiedemann (2002).

$$V_{t,s}^{EL} = \sum_{c \in SEC} p_{El,t,s}^{c} \cdot P_{c,t,s}^{EL} + p_{El,t,s}^{Spot} \cdot P_{Spot,t,s}^{EL} \tag{6-37}$$

$$V_{t,s}^{HT} = \sum_{c \in SHC} p_{HT,t,s}^{c} \cdot \dot{Q}_{t,s}^{HT} \tag{6-38}$$

$$V_{t,s}^{FU} = \sum_{f \in \mathbf{F}} \sum_{c \in SFC} p_{c,t,s}^{FU} \cdot P_{t,s}^{FU} \tag{6-39}$$

The total costs can similarly be divided into those linked to the power plant operation and those related to contracts. Furthermore they can be divided into two groups depending on whether they are influenced by decisions of the power plant operator (variable costs) or not (fixed costs). Hence, one can write:

$$C = \sum_{\tau=1}^{T} \sum_{s} \left[C_{PL,t,s}^{OP} + C_{PL,t,s}^{FIX} + C_{CT,t,s}^{OP} + C_{CT,t,s}^{FIX} \right] \tag{6-40}$$

with:

$C_{PL,t,s}^{OP}$ - total variable costs of unit operation (e.g. fuel costs, desulphurization ...)

$C_{PL,t,s}^{FIX}$ - total fixed costs of plant operation (e.g. man power, interests ...)

$C_{CT,t,s}^{OP}$ - total variable costs of contract exercise (e.g. energy price charged ...)

$C_{CT,t,s}^{FIX}$ - total fixed costs of contracts (e.g. base fee ...)

Among the variable costs, the operation costs (sum of start-up, continuous operation costs and shut-down costs for each unit within the power plant) and the variable costs for fuel contracts are included.

For the plant operation costs, one has to consider the costs of continuous operation $C_{OP,u,s,t}$ together with the start-up costs $C_{ST,u}$ and the shut-down costs $C_{SD,u}$, which are all linked to the corresponding binary unit operation variables:

$$C_{PL,var,s,t} = \sum_{u \in \mathbf{U}} \left[C_{ST,u} * U_{u,s,t} + C_{OP,u,s,t} * O_{u,s,t} + C_{SD,u} * D_{u,s,t} \right] \tag{6-41}$$

Thereby the start-up and shut-down costs $C_{ST,u}$ and $C_{SD,u}$ are taken as constant[31], whereas operation costs are a function of fuel price $p_{f,s,t}$ and fuel

---

[31] This clearly is a simplification, since especially start-up costs depend on the time, which the plant has cooled out since the last operation. The colder the plant, the more fuel is

input $\dot{Q}_{f,u,s,t}\left(P_{El,u,s,t}, P_{TH,u,s,t}\right)$, which is described, depending on the plant type, by Eqs. (6-5), (6-7), (6-22), (6-24), or (6-30):

$$C_{OP,u,s,t} = p_{f,s,t}\dot{Q}_{f,s,t}\left(P_{EL,u,s,t}, P_{TH,u,s,t}\right) \tag{6-42}$$

The variable contract costs include purchase costs for electricity, heat and fuel:

$$C_{CT,op}(t) = \sum_{c \in BEC} p_{EL,c}(t) \cdot P_{EL,c}(t) + \sum_{c \in BTC} p_{TH,c}(t) \cdot P_{TH,c}(t) + \sum_{c \in BFC} p_{f,c}(t) \cdot P_{f,c}(t) \tag{6-43}$$

The fixed plant and contract costs do not need to be considered if they are not influenced by the operator's decisions. They are then simply an additive constant in the optimization problem, which does not change the results.

Finally, the changes in stock correspond mostly to changes in the fuel storage during the optimization period and can be valued using:

$$\Delta S = \sum_{f \in F}\left[M_f(1) * p_f(1) - M_f(T) * p_f(T)\right] \tag{6-44}$$

with:

$M_{s,f}(1)$ - the amount of fuel f in the storage s at the beginning of the optimization period.

$M_{s,f}(T)$ - the amount of fuel f in the storage s at the end of the optimization period.

$p_f(1)$ - the market price for fuel f at the beginning of the optimization period.
$p_f(T)$ - the market price for fuel f at the end of the optimization period.

Since short time spans are considered, no interest or discount rates are applied within the optimization period.

2.3   Problem structure and possible simplifications

The equations described above yield a mixed integer linear programming problem. This problem is in general solvable if all parameters are known or at least assumed to be deterministic, yet the computational burden becomes

---

needed for getting it back to operation temperature (cf. also section 3.2). But this simplification is necessary to keep the problem formulation linear.

quite high even in the deterministic case when time horizons of several weeks or more are considered. Therefore, various decomposition approaches have been developed and applied to these types of problems mostly based on the principle of Lagrangian relaxation (cf. the discussion in Römisch 2001 and Huonker 2003).

The computational burden increases even further when it comes to optimization under uncertainty, so that the problem has to be scrutinized carefully in order to identify simplifications, which help to reduce the computational requirements.

A careful examination of the problem structure shows that the problem described above has the following, almost block diagonal structure (cf. also Huonker 2003):

$$
\max
\begin{bmatrix}
\mathbf{p}_{B1} - \mathbf{c}_{B1} \\
\mathbf{p}_{B2} - \mathbf{c}_{B2} \\
\dots \\
\mathbf{p}_{Bn} - \mathbf{c}_{Bn} \\
\mathbf{p}_{Spot}
\end{bmatrix}^{T}
\begin{bmatrix}
\mathbf{y}_{B1} \\
\mathbf{y}_{B2} \\
\dots \\
\mathbf{y}_{Bn} \\
\mathbf{y}_{Spot}
\end{bmatrix}
$$

$$
\therefore
\begin{bmatrix}
 & \mathbf{A}_{D,El} & & & \\
 & \mathbf{A}_{D,Th} & & & \\
 & \mathbf{A}_{Fuel} & & & \\
\mathbf{A}_{B1} & 0 & \dots & 0 & 0 \\
0 & \mathbf{A}_{B2} & \dots & 0 & 0 \\
\vdots & \vdots & \ddots & \vdots & \vdots \\
0 & 0 & 0 & \mathbf{A}_{Bn} & 0
\end{bmatrix}
\begin{bmatrix}
\mathbf{y}_{B1} \\
\mathbf{y}_{B2} \\
\dots \\
\mathbf{y}_{Bn} \\
\mathbf{y}_{Spot}
\end{bmatrix}
\leq
\begin{bmatrix}
\mathbf{b}_{D,El} \\
\mathbf{b}_{D,Th} \\
\mathbf{b}_{Fuel} \\
\mathbf{b}_{B1} \\
\mathbf{b}_{B2} \\
\dots \\
\mathbf{b}_{Bn}
\end{bmatrix}
\qquad (6\text{-}45)
$$

Thereby one block B with index k stands for one power plant or one contract. The equations for the different blocks of decision variables $\mathbf{y}_{Bk}$ are only coupled through the demand restrictions for electricity and heat as well as the fuel storage equation. The spot market variables $\mathbf{y}_{Spot}$ only enter the demand restrictions. The corresponding dual problem is then:

$$
\min
\begin{bmatrix}
\mathbf{b}_{D,El} \\
\mathbf{b}_{D,Th} \\
\mathbf{b}_{Fuel} \\
\mathbf{b}_{B1} \\
\mathbf{b}_{B2} \\
\cdots \\
\mathbf{b}_{Bn}
\end{bmatrix}^T
\begin{bmatrix}
\lambda_{D,El} \\
\lambda_{D,Th} \\
\lambda_{Fuel} \\
\lambda_{B1} \\
\lambda_{B2} \\
\cdots \\
\lambda_{Bn}
\end{bmatrix}
$$

$$
\therefore
\begin{bmatrix}
\mathbf{A}_{\mathbf{D},El}{}^T & \mathbf{A}_{\mathbf{D},Th}{}^T & \mathbf{A}_{Fuel}{}^T &
\begin{matrix}
\mathbf{A}_{B1}{}^T & 0 & 0 & 0 \\
0 & \mathbf{A}_{B2}{}^T & 0 & 0 \\
\vdots & \vdots & \ddots & \vdots \\
0 & 0 & 0 & \mathbf{A}_{Bn}{}^T \\
0 & 0 & 0 & 0
\end{matrix}
\end{bmatrix}
\begin{bmatrix}
\lambda_{D,El} \\
\lambda_{D,Th} \\
\lambda_{Fuel} \\
\lambda_{B1} \\
\lambda_{B2} \\
\cdots \\
\lambda_{Bn}
\end{bmatrix}
\geq
\begin{bmatrix}
\mathbf{p}_{B1} - \mathbf{c}_{B1} \\
\mathbf{p}_{B2} - \mathbf{c}_{B2} \\
\cdots \\
\mathbf{p}_{Bn} - \mathbf{c}_{Bn} \\
\mathbf{p}_{Spot}
\end{bmatrix}
$$

$$(6\text{-}46)$$

Here the particular structure of the matrices $\mathbf{A}_{\mathbf{D},El}$, $\mathbf{A}_{\mathbf{D},Th}$ and $\mathbf{A}_{Fuel}$ (cf. inequalities (6-32) to (6-34)) has to be considered: These matrices only contain the elements 0,1 and –1, since all energy flows are either added to or subtracted from the corresponding balance, and they have only one non-zero element per column, given that each energy flow at maximum enters one balance equation. The last (block) row of the dual problem then implies that

$$
\begin{aligned}
\lambda_{D,El} &\geq \mathbf{p}_{Spot,El} \\
\lambda_{Fuel} &\geq \mathbf{p}_{Spot,Fuel}
\end{aligned}
$$

$$(6\text{-}47)$$

and given that the dual problem is a minimization problem we can affirm that in the optimum the equalities

$$
\begin{aligned}
\lambda_{D,El} &= \mathbf{p}_{Spot,El} \\
\lambda_{Fuel} &= \mathbf{p}_{Spot,Fuel}
\end{aligned}
$$

$$(6\text{-}48)$$

hold. If we now integrate the first and the third block row of the original constraints into the original objective function using the Lagrange multipliers $\lambda_{D,El}$ and λ_{Fuel}, we obtain the new primal problem:

$$
\max \left(\begin{bmatrix} \mathbf{p}_{B1} - \mathbf{c}_{B1} \\ \mathbf{p}_{B2} - \mathbf{c}_{B2} \\ \dots \\ \mathbf{p}_{Bn} - \mathbf{c}_{Bn} \\ \mathbf{p}_{Spot} \end{bmatrix}^{T} - \mathbf{p}_{Spot,El}{}^{T} \mathbf{A}_{D,El}{}^{T} - \mathbf{p}_{Spot,Fuel}{}^{T} \mathbf{A}_{Fuel}{}^{T} \right) \begin{bmatrix} \mathbf{y}_{B1} \\ \mathbf{y}_{B2} \\ \dots \\ \mathbf{y}_{Bn} \\ \mathbf{y}_{Spot} \end{bmatrix}
$$

$$
\tag{6-49}
$$

$$
\ddots \begin{bmatrix} & & \mathbf{A}_{D,Th} & & \\ \mathbf{A}_{B1} & 0 & \dots & 0 & 0 \\ 0 & \mathbf{A}_{B2} & \dots & 0 & 0 \\ \vdots & \vdots & \ddots & \vdots & \vdots \\ 0 & 0 & 0 & \mathbf{A}_{Bn} & 0 \end{bmatrix} \begin{bmatrix} \mathbf{y}_{B1} \\ \mathbf{y}_{B2} \\ \dots \\ \mathbf{y}_{Bn} \\ \mathbf{y}_{Spot} \end{bmatrix} \leq \begin{bmatrix} \mathbf{b}_{D,Th} \\ \mathbf{b}_{B1} \\ \mathbf{b}_{B2} \\ \dots \\ \mathbf{b}_{Bn} \end{bmatrix}
$$

This problem has entirely a block structure, except for the heat demand block, which corresponds to Eq. (6-34). Consequently, the operation of any unit or contract, which does not deliver (or purchase) heat, can be optimized independently from all others[32]. This is the mark-to-market principle, which arises here directly from the structure of the optimization problem together with the assumption of unlimited purchases and sales to the spot market at uniform and constant prices. So this is an interesting and relevant special case, which will be looked at in somewhat more detail in the next section. Unfortunately, the assumption of unlimited purchases and sales to the spot market is even without CHP plants not fully valid in the continental European markets (cf. chapter 2). Once the day-ahead market is closed, the power plant operators are more or less set back to the traditional optimization with electricity demand (and possibly heat demand) restrictions, since they have only very limited possibilities left to sell or purchase on the market. So this case will be dealt with in section 4, taking into account the linkage between the next, demand restricted day and the following days, when electricity can be again sold and purchased on the market. The approach developed in this section is then extended in section 5 to the decision problem for longer-term portfolio management.

---

[32] Note that this holds only as long as risk neutrality is assumed (i.e. maximization of the expected value of the profit). For risk management, all power plants and other assets have to be valued together (cf. chapter 7).

3.   SEPARABLE OPTIMIZATION WITH
UNCERTAIN PRICES – REAL OPTION MODEL

The discussion in the previous section has shown that in the liberalized market, power plant operators can in principle operate each plant individually on a mark-to-market base. Given the price uncertainties on the market and the optionalities inherent in power plants, an attractive approach for analyzing this decision problem is to consider power plants as "real options". In fact, the real options approach is rooted in two different fields (cf. Ostertag et al. 2004). On the one hand, following notably the seminal book by Dixit and Pindyck (1994), investment under uncertainty has been analyzed, treating assets like power plants as (partly irreversible) investment opportunities. On the other hand, the theory and practice of financial markets has been extended to energy (and other commodity) markets and as a consequence the need arose to analyze the value and the risk of real assets like power plants in combination with traded products like forwards or futures. In this section, the focus is more on the latter approach to real option valuation, since it is more turned towards the operational decisions which are at the heart of this chapter. In section 3.1, first the basic models directly transposed from finance are briefly reviewed, then the more specific approaches developed in recent years are discussed in section 3.2. Finally, an approach is developed which explicitly copes with the market structure observed in continental Europe.

3.1   Basic models

A thermal power plant can be characterized most basically as a device capable of delivering at any given time electricity at a price equal to the variable costs for running the plant, i.e. mostly fuel costs. This corresponds in financial terms to an European call option, which provides to its holder a pay-off equal to the difference between the current price (of electricity) and the strike price of the option (the variable costs) (cf. e.g. Pilipovic 1998). If the price falls below the strike price of the option, the pay-off is zero, as shown in Figure 6-2. For the power plant this corresponds to shutting down the power plant instead of running it at loss.

Of course, most power plants are in reality not as flexible as to shut immediately down when prices fall below the threshold (cf. section 1) and therefore more realistic models are needed to describe power plant operation. These will be discussed in section 3.1. Beforehand, a few other points are however worth noticing:

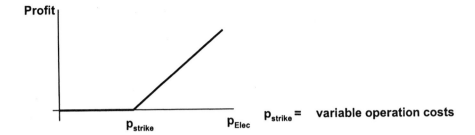

Figure 6-2. Pay-Off function for a thermal power plant, interpreted as European Call option

- The option of producing electricity from a power plant can be exercised repeatedly, every day or even each hour of a day. Thus, a thermal power plant has rather to be considered not as a single European option but as a sequence or "strip" of European options on different maturity dates.
- Hydro storage power plants are rather different from thermal power plants as far as their modeling in financial terms is concerned. Through the storage capacity, the operation statuses of the hydro plant at different moments in time are linked, the option therefore becomes path-dependent. If the storage capacity is small compared to the turbine capacity, the hydro storage plant can be considered as equivalent to an American option (option with exercise possibility at any time before maturity). Hydro pumping plants have again to be modeled differently, since they allow shifting electricity from low price to high price period. They have therefore characteristics similar to swap options, with low-price and high-price electricity being considered as two different commodities. But these options shall not be considered in detail here.
- Since electricity can hardly be stored, electricity delivered at different moments in time is in fact not the same commodity, but should rather be treated as different products (cf. section 4). This makes a strong difference to most other commodity prices and implies notably that methods for option valuation based on non-arbitrage arguments like the seminal Black and Scholes (1973) approach are not directly applicable.

Given that not only electricity prices are subject to uncertainty, but also fuel prices, power plants are more appropriately treated as a swap option (often called "spark-spread option"), that is an option to exchange one commodity (fuel) against another (electricity). Margrabe (1979) has provided here a valuation approach based on the Black-Scholes analysis, which has been widely applied in practice for valuing electric power plants (cf. e.g. Deng et al. 1998, Hsu 1998). Margrabe assumes geometric Brownian motion for the commodity prices (cf. chapter 4) and describes the value of the option on the basis of the price path of the swap (exchange ratio between the commodities). This ratio of the commodity prices follows also a geometric Brownian motion given the price processes of the commodities, so that the valuation problem is reduced to an option with one underlying, which follows a classical distribution. But, as seen before, the geometric Brownian motion is not a particularly good candidate for describing electricity price movements and furthermore the Margrabe model does not account for all the technical restrictions described in section 1.

3.2   Thermal power plants as path-dependent American options

The operation restrictions relevant in practice imply that the optimal operation at time *t* does not only depend on the price at time *t* but also on the prices in preceding and succeeding periods. The exercise of the option becomes hence (price-)path-dependent. In the last years, several models and solution methodologies have been developed to model these generalized real options, including those of Johnson et al. (1999), Gardner and Zhuang (2000), Tseng and Barz (2002), Deng and Oren (2002) and those in Ronn (2003).

The general approach of these models is rather similar although there are some small differences which of the restrictions discussed in section 1 (or even additional ones) are actually included. But all use a backward induction approach to determine the current value of the option starting from the set of possible values at some final exercise time. The backward induction is generally done using a stochastic dynamic programming approach but the methods differ in how the price scenarios are constructed. Gardner and Zhuang (2000) as well as Deng and Oren (2002) use a multinomial tree approach whereas Tseng and Barz (2002) use Monte-Carlo-simulations. The approach of Tseng and Barz (2002) is computationally intensive but rather flexible with regard to the price processes which can be modeled. Therefore it is used as a reference here.

3.2.1   Model of Tseng and Barz

Tseng and Barz (2002) start by formulating a quadratic cost function for the power plant, collapsing thus Eqs. (6-5) and (6-42) and extending (6-5) by including a quadratic term:

$$C_{Op,u,t} = p_{f,u}\dot{Q}_{f,u,t}\left(P_{El,u,t}\right) = p_{f,u}\left(a_0 O_{u,t} + a_1 P_{El,u,t} + a_2 P_{El,u,t}{}^2\right) \quad (6\text{-}50)$$

They then describe start-up costs, including their dependency on the previous shut-down time:

$$
C_{St,u,s,t} = \begin{cases} p_{f,u} a_{St,1,u}\left(1 - e^{O^{*}_{u,s,t-1}/\tau}\right) + a_{St,2,u} & \text{if } U_{s,t} = 1 \\ 0 & \text{if } U_{s,t} = 0 \end{cases}
\tag{6-51}
$$

The first term in the sum represents fuel costs for start up, whereas the second one captures other costs (e.g. labor). For incorporating the restrictions on minimum operation and minimum shut-down times, they introduce a state variable $O'_{u,t}$ (cf. Table 6-1).

Table 6-1. State variable in the real option model of power plant operation

Current value of state variable $O'_{u,t}$	Plant status	Possible values for the state variable $O'_{u,t+1}$ at the next time step	Corresponding values of the operation variables		
			$O_{u,t}$	$U_{u,t}$	$D_{u,t}$
$T_{Op,min}$	Plant is running since at least $T_{Op,min}$ hours	$T_{Op,min}$ or 1	1	0	0
$1 < O'_{u,t} < T_{Op,min}$	Plant is running since $O'_{u,t}$ hours	$O'_{u,t} + 1$	1	0	0
1	Plant is turned on	$O'_{u,t} + 1$	1	1	0
-1	Plant is turned off	$O'_{u,t} - 1$	0	0	1
$-1 < O'_{u,t} < -T_{cold}$	Plant is shut down since $-O'_{u,t}$ hours	$O'_{u,t} - 1$	0	0	0
$-T_{SD,min} \le O'_{u,t} < -T_{cold}$	Plant is now shut down since $-O'_{u,t}$ hours and can be turned on again	$O'_{u,t} - 1$ or 1	0	0	0
$-T_{cold}$	Plant is now shut down since at least T_{cold} hours	T_{cold} or 1	0	0	0

Source: own presentation based on Tseng and Barz (2002)

Positive values for $O'_{u,t}$ represent the case of an operating plant ($O_{u,s,t}=1$), with the value of $O^{*}_{u,s,t}$ describing the number of hours since start-up. Negative values conversely describe the case that the plant is shut down and the absolute value indicates the number of hours of shut down. Decisions on unit commitment are only possible in certain states: A start-up, i.e. $U_{u,s,t}= 1$ is only viable if $O^{*}_{u,s,t}$ previously at least reached the value $-T_{SD,min}$ and analogously a shut-down ($D_{u,s,t}= 1$) is feasible only if $O^{*}_{u,s,t}$ was previously equal to $T_{OP,min}$. Through these prespecified transition possibilities, transition between the different states is thus limited, so that only feasible operation sequences occur.

Additionally, Tseng and Barz (2002) account for maximum and minimum power constraints similar to (6-2) and (6-3) if the unit is turned on. If the prices are deterministic, the total profit to be maximized is written:

$$
G_{t_0 \to T}\left(\mathbf{O'}_{u \atop t \in [t_0\, T]}, \mathbf{P}_{EL,u \atop t \in [t_0\, T]}\right) = \sum_{t=t_0}^{T}\left(p_{EL,u,t}P_{EL,u,t} - p_{f,u,t}\left(a_0 O_{u,t}\left(O'_{u,t}\right) + a_1 P_{EL,u,t}\right.\right.
$$
$$
\left.\left. + a_2 P_{EL,u,t}{}^2\right) - C_{St,u,t}\left(O'_{u,t}\right)\right)
$$

$$(6\text{-}52)$$

This can also be written recursively:

$$
G_{t \to T}\left(\mathbf{O'}_{u \atop \tau \in [t\, T]}, \mathbf{P}_{EL,u \atop \tau \in [t\, T]}\right) = p_{EL,u,t}P_{EL,u,t} - p_{f,u,t}\left(a_0 O_{u,t}\left(O'_{u,t}\right) + a_1 P_{EL,u,t} + a_2 P_{EL,u,t}{}^2\right)
$$
$$
- C_{St,u,t}\left(O'_{u,t}\right) + G_{t+1 \to T}\left(\mathbf{O'}_{u \atop \tau \in [t+1\, T]}, \mathbf{P}_{EL,u \atop \tau \in [t+1\, T]}\right)
$$

$$(6\text{-}53)$$

The optimal profit at time t is then given by:

$$
G^{*}_{t \to T}\left(O'_{u,t_0-1}\right) = \max_{\substack{\mathbf{O'}_{u}, \mathbf{P}_{EL,u} \\ t \in [t_0\, T] \quad t \in [t_0\, T]}} G_{t_0 \to T}\left(\mathbf{O'}_{u \atop t \in [t_0\, T]}, \mathbf{P}_{EL,u \atop t \in [t_0\, T]}\right)
$$

$$(6\text{-}54)$$

It is thus depending on the value of the state variable at the time step preceding the optimization period. A recursive formulation is again possible and yields

$$
G^{*}_{t \to T}\left(O'_{u,t-1}\right) = \max_{O'_{u,t}}\left(p_{EL,u,t}P_{EL,u,t}\left(O'_{u,t}\right) - p_{f,u,t}\left(a_0 O_{u,t}\left(O'_{u,t}\right) + a_1 P_{EL,u,t}\left(O'_{u,t}\right)^2\right.\right.
$$
$$
\left.\left. + a_2 P_{EL,u,t}\left(O'_{u,t}\right)^2\right) - C_{St,u,t}\left(O'_{u,t}\right) + G^{*}_{t+1 \to T}\left(O'_{u,t}\right)\right)
$$

$$(6\text{-}55)$$

Thereby, the following relationship for the optimal output quantity $P_{EL,u,t}$ at given operation status $O_{u,t}$, is used:

$$
P_{EL,u,t}\left(O'_{u,t}\right) = O_{u,t}\left(O'_{u,t}\right)\min\left(P_{EL,u,\max}, \max\left(P_{EL,u,\min}, \frac{1}{2a_2}\left(\frac{p_{EL,t}}{p_{f,t}} - a_1\right)\right)\right)
$$

$$(6\text{-}56)$$

Applying Eq. (6-53), the problem can thus be solved recursively through dynamic programming. First, the optimal operation status at the final time T and the corresponding value of $G^*_{T \to T}(O'_{u,T-1})$ is determined for each value of $O'_{u,T-1}$. Then, for different values of the state variable $O'_{u,T-2}$ the optimal choice of $O'_{u,T-1}$ can be determined using the values $G^*_{T \to T}(O'_{u,T-1})$. This is then repeated backwards until the starting point t_0 is reached. Thereby one should note that for many values of $O'_{u,t-1}$ the optimization is in fact degenerated, since no alternative choices are possible given the transition restrictions described in Table 6-1. In fact an optimization is only possible if $O'_{u,t-1}$ equals $T_{OP,min}$, or if it is equal or below $-T_{SD,min}$. In the first case, the choice is between continued operation of the plant or turning it off. In the second case, the choice is inversely between continuation of the shut-down and start of the plant.

If prices are uncertain, the same basic principle can be used, however one has to account explicitly for the fact that the optimal profit does not only depend on the previous operation status, but also on the (now uncertain) prices.

Thus Eq. (6-53) has to be rewritten:

$$G_{t \to T}\left(\mathbf{O'}_{\substack{u \\ \tau \in [t\, T]}}, \mathbf{P}_{\substack{El,u \\ \tau \in [t\, T]}}, p_{El,t-1}, p_{f,t-1}\right) =$$

$$E\left[p_{El,t}P_{El,u,t} - p_{f,t}\left(a_0 O_{u,t}(O'_{u,t}) + a_1 P_{El,u,t} + a_2 P_{El,u,t}{}^2\right)\right.$$

$$\left. - C_{St,u,t}(O'_{u,t}) + G_{t+1 \to T}\left(\mathbf{O'}_{\substack{u \\ \tau \in [t+1\, T]}}, \mathbf{P}_{\substack{El,u \\ \tau \in [t+1\, T]}}, p_{El,t}, p_{f,t}\right)\middle| p_{El,t-1}, p_{f,t-1}\right]$$

$$(6\text{-}57)$$

Similarly the optimal profit is instead of (6-55) now given by:

$$G^*_{t \to T}(O'_{u,t-1}, p_{El,t-1}, p_{f,t-1}) = \max_{O'_{u,t}}\left(E\left[p_{El,t}P_{El,u,t}(O'_{u,t}) - p_{f,t}\left(a_0 O_{u,t}(O'_{u,t}) + \right.\right.\right.$$

$$\left.\left.\left. a_1 P_{El,u,t}(O'_{u,t}) + a_2 P_{El,u,t}(O'_{u,t})^2\right) - C_{St,u,t}(O'_{u,t}) + G^*_{t+1 \to T}(O'_{u,t}, p_{El,t}, p_{f,t})\right]\middle| p_{El,t-1}, p_{f,t-1}\right]$$

$$(6\text{-}58)$$

Hence in order to determine the optimal operation strategy $O'_{u,t}$ $(O'_{u,t-1}, p_{El,s,t-1}, p_{f,s,t-1})$ at time t, it is not only necessary to compute the pay-offs at time t in different price scenarios $p_{El,s,t}$ $p_{f,s,t}$, but also to determine the expected optimal profit at the further time steps given these different prices $p_{El,s,t}$ $p_{f,s,t}$. As before, the optimal operation strategy $O'_{u,t}$ $(O'_{u,t-1}, p_{El,s,t-1}, p_{f,s,t-1})$

is predetermined in many cases by $O'_{u,t-1}$ since requirements on minimum operation or minimum down times leave no choice for turn on/turn-off decision. So it is sufficient to look at those cases where $O'_{u,t-1} = T_{OP,min}$ or $O'_{u,t-1} \leq -T_{SD,min}$. For $O'_{u,t-1} = T_{OP,min}$, it is possible to determine a well-defined function $\tilde{p}_{El,t-1}(p_{f,t-1})$ for which the two possible strategies – continued operation or turn-off – yield the same pay off. For any price combination above $\tilde{p}_{El,t-1}(p_{f,t-1})$, it is optimal to continue the operation of the power plant, for all price combinations below $\tilde{p}_{El,t-1}(p_{f,t-1})$, it is optimal to turn off the plant. For the points on $\tilde{p}_{El,t-1}(p_{f,t-1})$, both alternatives are equally valuable, and it is therefore called indifference locus or turn-off barrier. Similarly, there exists an indifference locus $\tilde{\tilde{p}}_{El,t-1}(p_{f,t-1})$ for $O'_{u,t-1} \leq -T_{SD,min}$, called the turn-on barrier, since above these price combinations it is more profitable to turn on an offline plant, whereas below continued shut-down is preferable.

To determine the turn-on and turn-off barriers, Tseng and Barz propose to use Monte-Carlo simulations of future prices taking different price level $p_{El,s,t-1}$ $p_{f,s,t-1}$ as starting point. For each price path (scenario) and operation choice $O'_{u,t-1}$, the total pay-off is computed using the turn-off and turn-off barriers of higher time-steps to determine plant operation in this scenario. By comparing the different scenarios, the indifference loci $\tilde{p}_{El,t-1}(p_{f,t-1})$ and $\tilde{\tilde{p}}_{El,t-1}(p_{f,t-1})$ can then be determined. Hence, the optimal unit commitment problem may again be solved through backward induction in a dynamic programming approach. However, when moving one step backwards, it is always necessary to use Monte-Carlo simulations of price paths running until the final stage T, since the value of operation choice $O'_{u,t}$ is not only a function of the subsequent turn-on and turn-off decisions but also of $E\left[G^*_{t+1\rightarrow T}\left(O'_{u,t}, p_{El,t}, p_{f,t}\right)\middle| p_{El,t-1}, p_{f,t-1}\right]$ which is an average of a non-linear function of prices at stage t and at further stages.

The key assumptions for this model are that power output can be adjusted (within the limits given by $P_{EL,min,u}$ and $P_{EL,max,u}$) immediately to the current prices and that unit commitment decisions for time period t are done based on the price information available at time $t-1$. Thereby in the empirical applications the time periods are usually taken to be one hour (cf. Tseng and Barz 2002, Hlouskova et al. 2002). However, this is hardly realistic at least for continental European markets, where most trading takes place on a day-ahead basis.

Before proceeding further, two modifications to the methodology developed by Tseng and Barz (2002) are therefore proposed. The first one aims at reducing the computational burden whereas the second one aims at providing a more realistic model for markets with mostly day-ahead trading.

3.2.2 Use of lattices generated from Monte-Carlo simulations

As an alternative to the use of Monte-Carlo price paths, lattices (also called recombining trees or simply trees) are often used to model stochastic price developments (for real options in electricity markets, e.g. by Gardner and Zhuang 2000, Mo et al. 2000). With complex price models such as the one proposed in section 4, it is however hardly possible to determine analytically the appropriate values for the price steps and the corresponding transition probabilities. Hence, Monte-Carlo simulations are needed, but instead of performing them repeatedly at every time step, it is proposed to make one set of Monte-Carlo simulations starting from the initial time period t_0 and running until the final period T. The number n_{MC} of Monte-Carlo simulations should be chosen rather high, e.g. 2500 runs, in order to allow for sufficient accuracy in the next steps.

For each time step, the runs are then partitioned into n_G groups of similar scenarios, with n_G e.g. set equal to 50[33]. If there is only one stochastic variable[34], n_G groups of equal size $n_P = n_{MC}/n_G$ may be simply created by looking at the quantiles of the Monte-Carlo distribution. The sth group then corresponds to the n_P simulated values located between the quantiles $(s-1)/n_G$ and s/n_G (cf. Figure 6-3).

The group is then represented by one price $p_{s,t}$, which corresponds to the mean value of the simulations within the group. Of major importance are furthermore the transition probabilities $Tr_{s,t \to s',t+1}$ These correspond to the share of the Monte-Carlo simulations l, where the price at time t is within group s and the price at time $t+1$ is within the group s':

$$Tr_{s,t \to s',t+1} = \frac{card\left\{\left|p_{l,t} \in \left[p_{s,t}^{min}\ p_{s,t}^{max}\right] \wedge p_{l,t+1} \in \left[p_{s',t+1}^{min}\ p_{s',t+1}^{max}\right]\right|\right\}}{card\left\{\left|p_{l,t} \in \left[p_{s,t}^{min}\ p_{s,t}^{max}\right]\right|\right\}} \qquad (6\text{-}59)$$

---

[33] Alternatively a (recombining) tree may be constructed, where the number of nodes increases continuously for each time step. This may lower the computational burden if only short time intervals are considered. However, given the non-linearities in the pay-off function, this approach may induce additional inaccuracies. Also the non-arbitrage argument which motivates the tree construction in traditional path analysis in finance is not applicable here.

[34] I.e. the fluctuations of fuel prices are neglected, as also proposed by Gardner and Zhuang (2000)

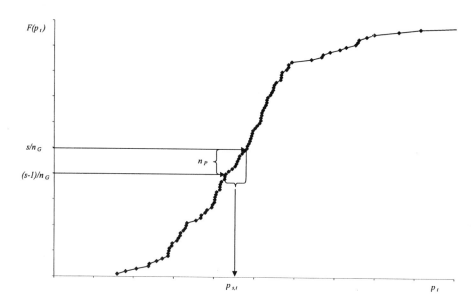

Figure 6-3. Construction of group prices from Monte-Carlo simulations

Using these prices and transition probabilities, Eq. (6-58) may be written in discretised version:

$$G^*_{t \to T}\left(O_{u,t-1}, P_{El,s,t-1}, P_{f,s,t-1}\right) = \max_{O_{u,s,t}} \left(\sum_{s'} Tr_{s,t-1 \to s',t} \left[P_{El,s',t} P_{El,u,s',t}\left(O_{u,s,t}\right) - P_{f,t}\left(a_0 O_{u,s,t}\left(O_{u,s,t}\right) \right. \right. \right.$$

$$\left. \left. \left. + a_1 P_{El,u,s,t}\left(O_{u,s,t}\right) + a_2 P_{El,u,s,t}\left(O_{u,s,t}\right)^2 \right) - C_{St,u,t}\left(O_{u,s,t}\right) + G^*_{t+1 \to T}\left(O_{u,s,t}, P_{El,s',t}, P_{f,s',t}\right) \right] \right)$$

$$(6\text{-}60)$$

This can now be recursively solved and if the value of $G^*_{t+1 \to T}$ is saved for all combinations of s' and $O'_{u,t}$, the evaluation for $t\text{-}1$ can be based solely on the use of the precomputed transition matrix $Tr_{s,t \to s',t+1}$ and of $G^*_{t+1 \to T}$, not requiring any new Monte Carlo simulations. This corresponds to working backwards through the lattice depicted in Figure 6-4.

For sufficient accuracy, both the number of price scenarios and the accuracy of the transition matrix have to be sufficient. But about 50 price scenarios should be largely sufficient, since at time intervals close to t_0, the price spread among the Monte-Carlo simulations will not be that large, and the impact of prices in the farer future – where price intervals may become larger - on the unit commitment decision at t_0 is more limited.

The approach as formulated here assumes that the prices follow a Markov process, i.e. the full information on future state probabilities is contained in the last observed prices, $p_{El,s,t-1}$ and $p_{f,s,t-1}$. The approach of Tseng and Barz can in principle also cope with more general price processes,

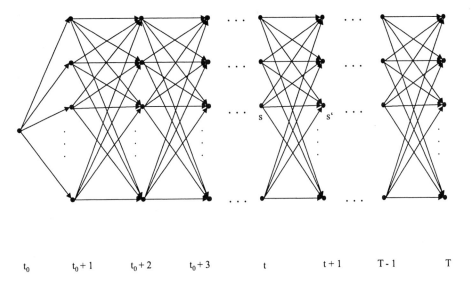

t_0 t_0+1 t_0+2 t_0+3 t $t+1$ $T-1$ T

Figure 6-4. Lattice for the real option model

but then already in Eqs. (6-57) and (6-58) the function value and the expectations have to be expressed not only as a function of the prices $p_{El,t-1}$ and $p_{f,\,t-1}$ at time $t-1$ but as a function of the whole set ψ_{t-1} of information available at time $t-1$. In this case, three- or more-dimensional matrices would be needed to describe the transition probabilities in the discretized version.

Another complication arises when there are several stochastic variables such as electricity and fuel prices. Then the simple grouping using the cumulative distribution function of the Monte-Carlo simulations does no longer work, since no unequivocal ordering of multivariate distributions is possible.

Instead a cluster analysis can be performed, using e.g. the k-means algorithm[35]. The distance function should thereby be chosen in a way to reflect the differences in profit which result from variations in the different stochastic variables. That is, the weights should be set equal to $\partial G^*_{t+1 \to T} / \partial p_{El,s,t-1}$ resp. $\partial G^*_{t+1 \to T} / \partial p_{f,s,t-1}$, evaluated at some appropriate value of the price variables. A first order approximation is to determine these

---

[35] A cluster analysis could also be used to avoid the multi-dimensional transition matrices in the case of non-standard Markov processes.

variations using the deterministic problem formulation and computing numerically the difference in profit induced by some small price increase, uniformly applied on all time steps[36]. As metrics the so-called city-block distance seems most appropriate, since under other metrics like the Euclidean or the squared Euclidean a few large clusters in the center of the distribution tend to be combined with many small clusters for more extreme values. In this case, both the mean and the transition probabilities for the small clusters may be very error-prone[37].

3.2.3   Adaptation to day-ahead-trading markets

Under the trading structures currently established in Germany and other continental European countries, the model of Tseng and Barz (2002) seems not to be adequate, since the assumed information and decision process does not correspond to the actual situation. Instead of modeling an hourly information arrival and scheduling problem a daily information structure should be considered. Thereby, two modeling approaches seem possible: on the one hand, an explicit modeling of the two stages of the daily trading process, namely the bid submission and the optimal scheduling once trades are closed and prices revealed. On the other hand, a common model encompassing the whole trading and scheduling process could be devised. The first approach will be looked at in section 4, therefore at this stage we are heading for a common model. For the basic choice on the link between information arrival and decision making, three alternatives are then possible:

1. The prices for the next day are revealed first, then the scheduling (unit commitment and production output) is done for the next day
2. The scheduling (unit commitment and production output) has to be done before the prices for the following day are known, only the prices of today can thus be used for making decisions for tomorrow
3. The unit commitment has to be done before the prices are known, but the production output can be chosen once prices have been revealed.

The third alternative obviously is in the spirit of the Tseng and Barz (2002) model (and also the model of Gardner and Zhuang (2000)), but it seems not very plausible in a day-ahead-trading setting. Also the second

---

[36] This is of course only a first approximation, since it only looks at marginal gains at average price values. Conceivable is an iterative scheme, updating the weights based on the results of the stochastic optimization. However further research is needed to clarify whether or under which conditions such a scheme is convergent.

[37] The least-squares Monte-Carlo method proposed by Longstaff and Schwartz (2001) could be another possibility to reduce the computational burden of the original Tseng and Barz approach.

alternative is not plausible as it assumes that the operator has to stick to his one-time chosen schedule independently of the prices revealed. The first alternative obviously is not entirely satisfactory either, but it can be well justified in the case that there is a continuous trading possible before the closure of the market. If this trading is efficient, prices and scheduling will evolve jointly and progressively until an equilibrium is reached, where the scheduling optimally fits the observed prices. Currently, continuous trading is done at the power exchanges in Germany and elsewhere only for standard base and peak products, yet through OTC trading also some additional market efficiency and convergence is reached. Thus, the first alternative, albeit not fully satisfactory either, is certainly the most preferable.

In this case, the recursive formulation of the optimal choice problem analogous to Eq. (6-58) is:

$$G^*_{d \to d_{max}}\left(O'_{u,d-1,24}, \mathbf{p}_{El,d}, p_{f,d}\right) = \max_{O'_{u,d}}\left(\sum_{h=1}^{24}\left(p_{El,d,h}P_{El,u,d,h}\left(O'_{u,d,h}\right) - \right.\right.$$

$$p_{f,d}\left(a_0 O_{u,d,h}\left(O'_{u,d,h}\right) + a_1 P_{El,u,d,h}\left(O'_{u,d,h}\right) + a_2 P_{El,u,d,h}\left(O'_{u,d,h}\right)^2\right) - \quad (6\text{-}61)$$

$$\left. C_{St,u,d,h}\left(O'_{u,d,h}\right)\right) + E\left[G^*_{d+1 \to d_{max}}\left(O'_{u,d,24}, \mathbf{p}_{El,d+1}, p_{f,d+1}\right)\middle|\mathbf{p}_{El,d}, p_{f,d}\right]\right)$$

A double time indexing for days d and hours h is used here and the electricity prices for one day are grouped in the vector $\mathbf{p}_{El,d}$ to illustrate the double time scale of the decision problem. For fuel prices $p_{f,d}$ no intraday variations are assumed given the storage possibilities. The optimal gain function is then dependent on the operation state in the last hour of the previous day $O'_{u,d-1,24}$ and it is only stochastic in as far as the decisions made for day d have an impact on the future (uncertain) profits summarized in $G^*_{d+1 \to dmax}$. The corresponding expected values, labeled $\psi(O'_{u,d,24} \mid \mathbf{p}_{El,s,d}, p_{f,s,d})$, may again be determined using a transition matrix approach similar to (6-60):

$$\psi\left(O'_{u,s,d,24}\middle|\mathbf{p}_{El,s,d}, p_{f,s,d}\right) = E\left[G^*_{d+1 \to d_{max}}\left(O'_{u,s,d,24}, \mathbf{p}_{El,d+1}, p_{f,d+1}\right)\middle|\mathbf{p}_{El,s,d}, p_{f,s,d}\right]$$

$$= \sum_{s'} T_{s,d \to s',d+1} G^*_{d+1 \to d_{max}}\left(O'_{u,s,d,24}, \mathbf{p}_{El,s',d+1}, p_{f,s',d+1}\right)$$

$$(6\text{-}62)$$

When the optimization problem is solved through dynamic programming, it is hence possible first to compute the expected value $\psi(O'_{u,d,24} \mid \mathbf{p}_{El,s,d}, p_{f,s,d})$

for each final operation state $O'_{u,d,24}$. It may thus be interpreted as the shadow value for state $O'_{u,d,24}$ by the end of the planning day. Then the planning for the day may be done using either deterministic dynamic programming as discussed in the context of Eq. (6-55), or by using non-linear mixed integer programming to obtain the optimal schedule for all hours at once.

4.   TWO-STAGE OPTIMIZATION PROBLEM IN DAY-AHEAD MARKETS INCLUDING CHP

As described earlier, the optimization in the day-ahead market is done in two steps. First, the optimal bids to be submitted on the market are computed, and then the optimal schedule is determined knowing how much energy has been sold to the spot market. This means that the planning at the second stage is done with a load constraint to be fulfilled, as was the case before liberalization. But the terminal conditions for this load constrained unit commitment and load dispatch are different: after the end of the planning horizon, there is no binding load constraint, since the trade for the following period has not yet closed. For a conventional power plant system without CHP, the final unit commitment status of each power plant should hence be assigned a monetary value equal to the value obtained from the real option approach.

This will lead to an optimal scheduling of the power plants since the difference in the option values for various plant statuses precisely reflects the value for the power plant of being turned on or off, depending on the price expectations for the following day and the resulting earning possibilities in the subsequent days.

For a CHP plant, there is no such obvious value of being turned on or off, since this value will depend both on market prices and on the heat demand in the following days. The value of being turned on or off can, if at all, only be determined through a stochastic optimization of the following planning period(s). Hence a logical approach is to do a two-stage stochastic optimization with given electricity and heat demand for the first stage (day *T*) and various scenarios on heat demand and electricity prices for the second stage (day *T*+1) (cf. Figure 6-5). The objective function for this problem may be derived from Eqs. (6-35) ff. Under the separability assumption (cf. section 2.3), only sales to the spot market and variable production costs have to be considered. Applying furthermore the two-stage structure of the problem, the following formulation is obtained by collapsing the Eqs. (6-35), (6-36), (6-37), (6-40), (6-41) and (6-42):

$$\max_{\dot{Q}_{f,u,t,s},P_{EL,u,t,s},P_{TH,u,t,s},U_{u,t,s},P^{EL}_{Spot,t,s}} G_T$$

$$\therefore G_T = \sum_{t\in[t_0,t_1-1]}\left[-\sum_{u\in U}\left[C_{ST,u}*U_{u,t}+p_{f,u}\dot{Q}_{f,u,t}\left(P_{EL,u,t},P_{TH,u,t}\right)\right]\right]+$$

$$\sum_{s}\mathrm{Pr}_s\sum_{t\in[t_1,t_2-1]}\left[p^{Spot}_{El,t,s}\cdot P^{EL}_{Spot,t,s}-\sum_{u\in U}\left[C_{ST,u}*U_{u,s,t}+p_{f,s}\dot{Q}_{f,u,s,t}\left(P_{EL,u,s,t},P_{TH,u,s,t}\right)\right]\right]$$

$$(6\text{-}63)$$

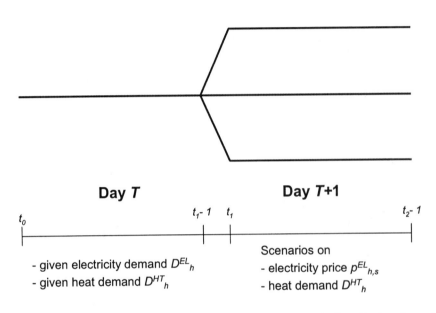

Figure 6-5. Two stage optimization problem for CHP operation in liberalized markets with daily auctions

The first part of the objective function, corresponding to the hours t_0 to t_1- 1, represents day T where the revenues have already been fixed, so only costs are optimized. Furthermore different scenarios need not to be considered[38]; therefore the scenario index s has been suppressed for notational convenience. For day $T+1$ with hours t_1 to t_2-1, the quantities $P^{EL}_{Spot,t,s}$ sold to the spot market are also subject to optimization and prices as

---

[38] Heat demand could be treated as uncertain also on the first day, but this effect is neglected here. Forced plant outages constitute a further source of uncertainty, but since they are rather infrequent they are not included in the daily planning process but are handled when they occur.

well as heat demand are uncertain. The optimization is done under the operation restrictions of the different power plants (cf. section 1) and with the supply and demand balances for power and heat valid at each hour (cf. section 2.1):

$$\sum_{u \in EU} P_{u,t,s}^{EL} \geq P_{CT,t,s}^{EL} + P_{Spot,t,s}^{EL} \tag{6-64}$$

$$\sum_{u \in HU_i} \dot{Q}_{u,t,s}^{HT} \geq \dot{Q}_{CT,t,s}^{HT} \tag{6-65}$$

All sales and purchases on a contract basis, be them with final customers or on the forward/futures market, are thereby lumped together into the quantities $P_{CT,t,s}^{EL}$ and $\dot{Q}_{CT,t,s}^{HT}$, considering that these have to be treated as exogenous for operation planning purposes. Of course, if contracts with load flexibility have been negotiated with customers, these should also be optimized in the operation scheduling.

An open question is then how to set adequate terminal restrictions at the end of day two. Given that the planning horizon is in fact the first day and that the second day is only included for improving the results for the first day, the termination effects at the end of day two should in general be of minor importance. If this is not the case (e.g. power plant with very high start up costs), the approach described in the following section for longer-term optimization may be applied in the same way.

5.   LONGER-TERM PORTFOLIO MANAGEMENT

Given the volatility of spot market prices, a longer-term portfolio management, which limits the impact of the day-to-day fluctuations while maintaining the profit opportunities, is of high interest to the power plant operators. Forward and futures markets offer good opportunities for building up a portfolio which is not as much exposed to price risks as if power plants are only submitted as bids on the day-ahead market. Of course the question then arises how much should be sold optimally on the forward markets[39].

---

[39]   In the following, often the short expression "forward market" is used for designating both forward and future markets (cf. also Mo et al. 2000). Since we will not focus on the financial and delivery details of the products traded in these markets, both are equivalent

Other studies which have addressed this is issue of longer term portfolio selection include Fleten and Wallace (1998), Mo et al. (1999), Fleten et al. (2002) and Bagemihl (2002). Except the last one, these authors focus however on portfolios with only hydropower generation. Bagemihl (2002) deals with a combined hydro-thermal generation park, but only in a deterministic setting.

If there is any difference between the (discounted) expected spot market price and the current forward price, then the power plant operator has obviously an incentive to sell as much as possible on the market with the higher price – at least if he or she is risk neutral. Since a similar argument holds in fact for any actor (trader, retailer) operating on both markets, the presence of a number of risk neutral actors will be sufficient to make spot markets and forward markets converge in expectation, i.e. formally:

$$p_{FW,T_0,T}^{EL} = E\left[\bar{p}_{Spot,T}^{EL}\right] \tag{6-66}$$

Given the non-storability of electricity, cost of carry or convenience yield arguments applied for other commodities (cf. e.g. Pindyck 2001) are not applicable to electricity (cf. e.g. Botterud et al. 2002). Also "no arbitrage" arguments linking the forward price to the current spot price are not applicable given the non-storability, but under risk neutrality the arbitrage possibilities between selling now or selling later (respectively buying now or buying later) imply the convergence in expectation of Eq. (6-66).

If no risk neutrality for the market as a whole can be assumed, Eq. (6-66) has to be modified slightly, including the market price of risk ρ, which is the difference between the applicable interest rate in the market ρ_{EL} and the risk free interest rate ρ_0 (cf. e.g. Botterud et al. 2002):

$$p_{FW,T_0,T}^{EL} = e^{-\rho(T-T_0)}E\left[\bar{p}_{Spot,T}^{EL}\right] \tag{6-67}$$

Whether this market price of risk is positive or negative depends on the relative importance of risk aversion from sellers and buyers of electricity on the wholesale market. If the risk aversion of buyers dominates, the required interest payments for forward contracts will be below the interests gained from risk-free assets, i.e. the market price of risk will be negatively. Conversely if the risk aversion of sellers dominates, the market price of risk will be positive. Botterud et al. (2002) found for the Nordpool market that

---

for our considerations and they are subsumed under the term forward market, since this market is more liquid in Germany and most continental European markets.

the market price of risk is significantly below zero, providing evidence that the risk-aversion of buyers dominates over the risk-aversion of sellers. This can be explained by the fact that the price risk for buyers is unlimited (extreme price spikes may occur) whereas prices are generally bounded by zero from below, so that the price risk for sellers is limited.

If the risk preferences of the power plant operator correspond to those of the market, he will be indifferent to selling at the forward or at the spot market. But even if the risk preferences differ from the average, it is preferable to separate the valuation of the portfolio at market prices from those valuations and trades, which are a consequence of deviations in risk valuations. Under the former, earnings can be transferred from the spot to the future market without changing the (market) value of these earnings whereas the latter creates additional utility for the operator. Either his risk exposition is further reduced through additional forward sales if he is more risk adverse than the market or he increases his expected earnings and his individual utility through risk overtaking from other market participants.

In the following, the focus is on the market valuation of a given power plant portfolio as a basis for an adequate portfolio management whereas risk issues are dealt with in chapter 7. In order to obtain the market value of a power plant portfolio, we have to determine its optimal exercise strategy over the whole delivery period, taking into account the stochastics of prices and other uncertain factors.

In the case of ordinary power plants without CHP, the real options approach described in section 3 may directly be extended to cover the whole delivery period. Of course, this requires the use of efficient computational techniques such as those proposed in section 3.2.2.

One should note that this valuation methodology will provide rather different results than a valuation based on price models fitted to the forward curve such as those proposed by Clewlow and Strickland (2000). In fact, the spot price $\overline{p}_{Spot,T}^{EL}$ in Eqs. (6-66) and (6-67) is an average spot price over all hours corresponding to the forward period T:

$$\overline{p}_{Spot,T}^{EL} = \frac{1}{card(\mathbf{H}_T)} \sum_{t_D \in \mathbf{H}_T} p_{Spot,t_D}^{EL} \qquad (6\text{-}68)$$

In the real option model, the power plant is treated as a strip of (interconnected) options for the different hours t_D, whereas in any approach based on forward curve price models, the valuation of a power plant would be as a single option for the whole forward period. This is clearly not very satisfactory given the systematic daily and weekly variation in hourly spot prices, which imply a corresponding variation in exercise probability for the power plant option.

For CHP systems, the real options approach is not directly applicable as explained before. Yet an approach is needed, which considers all the nodes in the price (and heat demand) lattice depicted in Figure 6-4. One possibility is to apply the approach developed in section 4 to each node in the lattice for the planning period, weighting the different nodes with their respective probabilities. But the drawback is then that the links between the different planning periods are only taken into account within the two-days planning horizon. Beyond the planning horizon, no interdependencies are considered.

To overcome this problem, the scheme depicted in Figure 6-6 is proposed. Still two-days optimizations are carried out for each node in the lattice, but the terminal conditions in the optimization problem are modified to include information available from the optimization of the following time steps. Thus the scheme proceeds by backward induction as in dynamic programming or in the real option approach discussed above.

The information to be transferred within this backward induction approach is the one contained in the equations linking several time steps together. These equations are in the following called dynamic equations. In the system under study, the only dynamic equations are the start-up and shut-down Eqs. (6-9) and (6-11) to (6-13) as well as the minimum up and the minimum down time requirements formulated in Eqs. (6-13) and (6-14). For these restrictions, it is either possible to use the primal or the dual variables for the transfer of information. For the start-up and shut-down equations, the use of the dual variables seems appropriate since these reflect the opportunity value of having the unit running (for the start-up equations) respectively having it turned off. Equation (6-9), which is recalled here for convenience

$$U_{u,t,s''} \geq O_{u,t,s''} - O_{u,t-1,s'} \qquad\qquad \forall S \; \forall s'', s' \in S \qquad\qquad (6\text{-}9')$$

implies for example, if it is binding, a shadow price $\psi_{u,t,s''}^{ST1}$. If now the time step $t=t_2$ is outside the planning horizon (because it is at the beginning of day $T+2$), the shadow price $\psi_{u,t,s''}^{ST1}$ may be applied to the decision variable $O_{u,t2-1,s'}$ which is still within the planning horizon. Hence the objective function has to be extended by a term

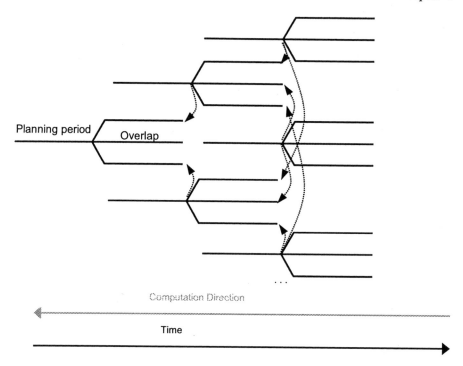

Figure 6-6. Backward stochastic induction for CHP power plants

$$L_{u,t_2,s''} = (-1) \cdot \psi_{u,t_2,s''}^{ST1} \cdot \left(-O_{u,t_2-1,s'}\right) \qquad \forall S \ \forall s'',s' \in S \quad (6\text{-}69)$$

This corresponds to writing the unconstrained Lagrangian function instead of the constrained objective function, at least with respect to this time-overwhelming restriction. The premultiplier (-1) in the Lagrangian is to conform to Lagrangian functions for LP problems (cf. e.g. Vanderbei 2001, p. 282). The factor $(-O_{u,t2-1,s'})$ is exactly what appears in Eq. (6-9) when it is brought to the standard form $\mathbf{Ax} \leq \mathbf{b}$. Since several states s'' may follow the state s', corresponding to different scenarios S, the Lagrangian adder has to be formulated for each of these scenarios. These adders may however be summed up, yielding the overall adder:

$$\tilde{L}_{u,t_2-1,s'} = \left(\sum_{\substack{s'' \\ (s'',s') \in S}} \psi_{u,t_2,s''}^{ST1} \right) \cdot O_{u,t_2-1,s'} \qquad \forall S \ \forall s'',s' \in S \quad (6\text{-}70)$$

Analogous extensions to the objective function have to be formulated to cope with the start-up and shut-down equations (6-11) to (6-13).

The objective function for the optimization problem at day T in node (or state of the world) s may then be formulated as:

$$
G_{T,s} = \sum_{t \in [t_0, t_1]} \left[\sum_{u \in U} \left[p_{El,s,t}^{Spot} \cdot P_{El,u,s,t} - C_{ST,u} * U_{u,s,t} - p_{f,u,s,t} \dot{Q}_{f,u,s,t} \left(P_{El,u,s,t}, P_{TH,u,s,t} \right) \right] \right] +
$$

$$
\sum_{s'} Tr_{T,s \to T+1,s'} \sum_{t \in [t_1, t_2 - 1]} \left[\sum_{u \in U} \left[p_{El,s',t}^{Spot} \cdot P_{El,u,s',t} - C_{ST,u} * U_{u,s',t} + p_{f,s',t} \dot{Q}_{f,u,s',t} \left(P_{El,u,s',t}, P_{TH,u,s',t} \right) \right] \right]
$$

$$
\sum_{s'} Tr_{T,s \to T+1,s'} \sum_{u \in U} \left(\sum_{\substack{s'' \\ (s'',s') \in S}} \left(\psi_{u,t_2,s''}^{ST1} - \psi_{u,t_2,s''}^{ST3} - \psi_{u,t_2,s''}^{SD1} + \psi_{u,t_2,s''}^{SD2} \right) \right) \cdot O_{u,t_2-1,s'}
$$

$$(6-71)$$

It corresponds thus to the objective function of the short-term optimization problem described in section 4 except for the third term, which describes the value of having the unit u turned on at the end of day $T+1$. Another difference is that now the results of the spot market auctions for day T are not yet known, so that sales to the spot market are in principle an additional decision variable. On the other hand, the electricity sold to contracts should also be valued according to the mark-to-market principle. In the end, this means that all electricity produced during days T and $T+1$ should be priced with the corresponding spot market price. It can hence directly be included in the objective function and the electricity balance equation can be dropped.

For the minimum up and minimum down time restrictions, an information transfer using the primal variable is more appropriate, since the reduced two-day problem should still lead to operation schedules which are feasible in the larger multi-day setting. The original restriction (6-13) is therefore split up in those parts which are within the reduced problem and those which are outside:

$$
U_{u,t,s'} T_{u,\min}^{UP} \leq \sum_{\tau=t}^{t_2-1} O_{u,\tau,s'} + \sum_{\tau=t_2}^{t+T_{u,\min}^{UP}-1} O_{u,\tau,s''} \qquad \forall S \ \forall s', s'' \in S \quad (6-72)
$$

The second sum includes only decision variables which are outside the scope of the two-day planning horizon. Therefore, these should be replaced by their optimal values obtained in the next overlapping problem. Given that several states s'' may follow the state s' after the end of day $T+1$, the restriction has to be valid for each of them. The hardest restriction is the one with the minimum operation during day $T+2$. Equation (6-72) is thus reformulated:

$$U_{u,t,s}\, T^{UP}_{u,\min} \le \sum_{\tau=t}^{t_{\max}} O_{u,\tau,s} + M_{u,t,s} \qquad\qquad \forall s \qquad\qquad (6\text{-}73)$$

with $M_{u,t,s}$ representing the minimum operation during the beginning of day $T+2$:

$$M_{u,t,s} = \min_{s'}\left(\sum_{\tau=t_{\max}+1}^{t+T^{UP}_{u,\min}-1} O_{u,\tau,s'} \right) \qquad\qquad (6\text{-}74)$$

$M_{u,t,s}$ is indexed over t, since it includes only the operation states $O_{u,\tau,s'}$ of those time steps τ which are before the end of the minimum up time as viewed from time t, which is $t + T^{UP}_{u,\min} - 1$.

Similarly to Eq. (6-73), the restrictions for the minimum down time have to be adapted for those time steps t, which are close to the end of the planning horizon.

The proposed solution methodology is not an exact decomposition, since there is no master problem, which would be repeatedly used for updating the dual or primal variables. It is neither a full-fledged dynamic programming approach, since no complete state description of the possible states of all units in the system is used for backward induction. It is rather an approximation methodology which takes exactly into account that decisions at any day T are dependent on the uncertainties concerning the states of the world on day $T+1$, but which only approximately includes the dependency on uncertainties in the farer future. This approach is consequently best suited for systems where the strength of interconnections is rapidly declining with increasing time lag and it is less suited for systems with strong long-term interconnections. In the CHP system considered, the only direct link between non-adjacent time steps is through the equations for minimum up and minimum down times. Furthermore the approximation should be well suited if there is no cumulative variable such as the reservoir level in hydro systems. The use of the proposed approximation seems therefore adequate even if no formal proof for the adequacy can be given.

As a result of this solution methodology, the value of the portfolio of CHP plants is obtained as well as the expected operation strategies and the variations of these operation strategies depending on prices and heat demand. Also the expected electricity production quantities and the corresponding fuel consumptions are obtained, which can then be used for hedging purposes. If changes in forward prices directly translate into changes in the expected hourly spot prices under an additive price model (cf. chapter 7, section 4), then selling the expected electricity production on the forward market and purchasing the expected fuel quantities on the fuel markets will eliminate the short-term price risks, since decreases in the value of the power plant (e.g. through lower electricity prices) will be offset by decreases in the value of the electricity sold and vice versa, making the hedged portfolio risk free. Hence this strategy of "expected operation hedging" corresponds in financial terms to establishing a "delta hedge" eliminating in the short term the electricity and fuel price risks (cf. Pilipovic 1998, Bergschneider et al. 1999, Hull 2000). Yet this risk freeness is only valid under specific conditions and, if at all, only applicable in the short term on the forward market. But these issues will be discussed in detail in chapter 7.

6. APPLICATION

The methods developed in sections 3 to 5 are in the following applied to a combined portfolio of several power plants. It includes single power plants for which the impact of various parameter variations on the results can be analyzed in detail and also a more complex CHP system, which is close to real systems (cf. section 6.1). For these portfolios, the real options approach is applied in section 6.2, the two-stage short-term optimization in section 6.3 and the longer-term optimization in section 6.4.

6.1 Portfolio description

The overall portfolio considered is composed of two non-CHP power plants (cf. section 6.1.1) and a CHP system consisting of two interconnected district heating grids with a total of eight power plants (cf. section 6.1.2). The characteristics of the power plants without cogeneration correspond to typical power plants built during the last decade in Germany whereas the CHP system is rather similar to part of the system operated by the Berlin utility BEWAG.

6.1.1 Thermal power plants without cogeneration

Two different plant types, a gas-fired and a coal-fired power plant, are considered in order to illustrate the impact of various plant parameters. The data for these plants are summarized in Table 6-2. The fuel prices for the power plants have to be deduced from the import prices described by the fuel price model of section 4.3.1 adjusted by the corresponding transport costs. The operation restrictions are only considered for the coal-fired plants, since gas plants are rather flexible.

Table 6-2. Power plants without cogeneration considered for application case

	Coal-fired power plant	Gas-fired combined cycle power plant
Maximum net output $P_{u,max}^{EL}$ [MW]	600	600
Minimum net output $P_{u,min}^{EL}$ [MW]	250	250
Fuel	Coal	Gas
Marginal fuel consumption β_u^{EL}	2.10	1.70
Additional fuel consumption at minimum output β_u^O [MW]	80	60
Fuel price		
Import prices [€/MWh]	5.0	11.0
Transport costs [€/MWh]	0.8	1.4
Price at power plant $p_{f,s,t}$	Import price + Transport cost	
Other variable costs [€/MWh$_{el}$]	2.2	1.2
Operation restrictions considered		
Minimum operation time [h]	8	1
Minimum shut down time [h]	8	1
Start up costs $C_{St,u}$ [€]	20,000	30,000

6.1.2   Cogeneration system

The whole system consists of eight CHP plants and two boilers with two main district heating grids (Figure 6-7). Six out of the eight CHP plants are coal fired plants (turbines T3, T4, T5, T6, T7, T8) with extraction condensing turbines; the other two are gas turbines. The main characteristics of these plants are summarized in Table 6-3.

The heat demand is given as hourly demand curve for a reference year. Figure 6-8 illustrates how heat demand varies between seasons but also within one season.

Variations in heat demand are modeled through an ARMA model with temperature as exogenous variable. For temperature, an ARMA model is specified and used for forecasting. Thereby the seasonal effects are approximated through sinus and cosinus variables (cf. chapter 4, section 2.2).

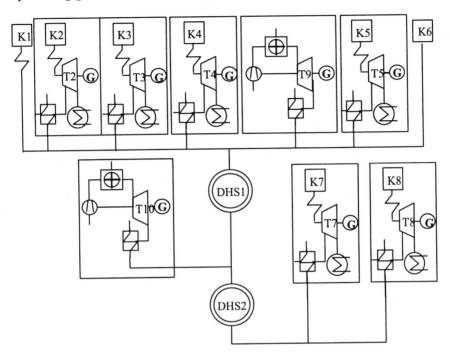

Figure 6-7. Schematic representation of CHP system considered for application case

6.2   Real options approach

In a first step, the real option approach developed in section 3.2 is applied to the two power plants without cogeneration described in section 6.1.1. As price model, the integrated model developed in chapter 4 is used. The valuation is done for the months of January, August and October 2002, based on the data available at the end of the previous months.

For the price model, 2500 Monte-Carlo simulations are carried out and the power plants are valued using the approach described in section 3.2.3. As a result, the optimal operation of the plants is obtained and their option values (cf. Table 6-3). Given that fuel costs are much lower for the coal-fired plant than for the gas-fired plant, the option value of the coal plant is considerably higher in all months. But the value of the gas plant is a great deal higher than the time value, which is obtained when using the expected values of the spot market prices for valuation.

Table 6-3. Cogeneration power plants considered for case study

Extraction-condensing steam turbines

	K2/T2	K3/T3	K4/T4	K5/T5	K7/T7	K8/T8
Maximum electricity output $P_{u,max}^{EL}$ [MW]	123.5	284.0	284.0	90.0	142.5	142.5
Maximum heat output $P_{u,max}^{TH}$ [MW]	183.0	370.0	370.0	136.0	204.0	204.0
Fuel	Coal	Coal	Coal	Coal	Gas	Gas
Minimum fuel consumption β_u^O [MW]	3.00	41.0	31.9	20.6	15.45	45
Marginal fuel cons. for electricity β_u^{EL}	2.81	2.40	2.45	2.40	2.40	2.40
Marginal elec. loss for heat production $\beta_u^{HT}/\beta_u^{EL}$	-0.11	-0.13	-0.13	-0.12	-0.152	-0.152
Minimum electricity production $P_{u,min}^{EL}$ [MW]	55.0	140.0	140.0	52.0	38.0	38.0
Minimum electricity to heat ratio $s_{u,min}^{EL}$	0.56	0.64	0.64	0.54	0.54	0.54
Minimum operation time $T_{u,min}^{UP}$ [h]	24	12	12	48	8	8
Minimum down time $T_{u,min}^{DOWN}$ [h]	24	12	12	24	4	4

Gas turbines   *Boilers*

	T9	T10		K1	K6
Maximum electricity output $P_{u,max}^{EL}$ [MW]	82.8	77.7			
Maximum heat output $P_{u,max}^{TH}$ [MW]	95.0	145.0	$P_{u,max}^{TH}$ [MW]	93	106.5
Fuel	Gas	Oil	Fuel	Oil	Oil
Minimum electricity production $P_{u,min}^{EL}$ [MW]	23.0	23.0			
Minimum fuel consumption β_u^O [MW]	0	51.4	Efficiency η [%]	91.2	92
Minimum electricity to heat ratio β_u^{EL}	3.6	2.6			

Figure 6-9 and Figure 6-10 show the impact of the operation restrictions on the option value of the power plant. For the coal-fired plant, the reduction of minimum operation and minimum shut-down times from 12 to 0 hours leads only to an increase by about 0.2 % in the option value, given that the power plant is in most hours deep "in the money", i.e. the hourly price

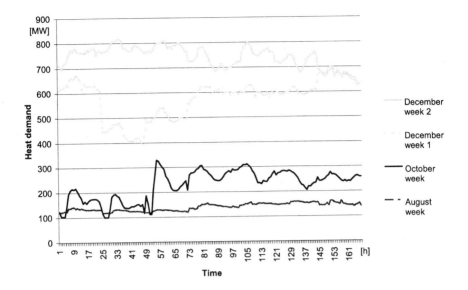

Figure 6-8. Selected heat demand curves for application case

Table 6-4. Option value of the power plants analyzed

	Coal-fired plant	Gas-fired plant
January 2002	5.49 M€	2.07 M€
April 2002	3.44 M€	1.42 M€
July 2002	3.60 M€	1.76 M€
October 2002	2.89 M€	0.63 M€

Source: own calculations, based on market price simulations with the integral model from chapter 4, using information available at the end of the preceding month (no correction for forward quotes made)

exceeds by far the operating costs of the power plant, so that it will be run continuously and operation restrictions have little effect. Furthermore the start-up costs make a short-time operation anyhow unattractive, so that the restriction on operation and shut-down time do not affect much the actual operation.

The impact of reduced start-up costs is also most significant in the case of the gas-fired plant, since it will be more frequently turned on and off given its higher variable costs (cf. Figure 6-10). In fact the effect of reduced start-up costs is double: on the one hand each start-up effectuated in the optimal operation scheme is less penalized and furthermore additional start-ups are carried out, leading to additional profits. Overall, the reduction of start-up costs from 30,000 € to zero would increase the earnings of the gas-fired plant by approximately 87 % while the earnings of the coal-fired plant

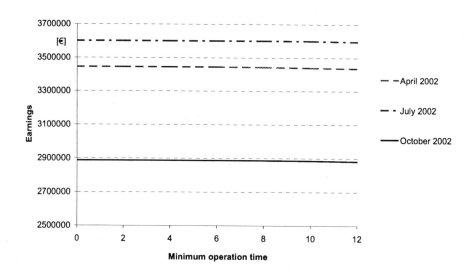

Source: own calculations, market price simulations as in Table 6-4

Figure 6-9. Option value of the coal-fired power plant depending on the operation time restrictions

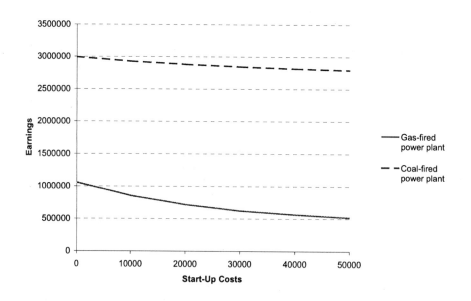

Source: own calculations, market price simulations for October 2002 as in Table 6-4

Figure 6-10. Option value of the power plants analyzed depending on the start-up costs

are only augmented by about 6 %. These results show that a modeling of power plants as real options is especially relevant for options, which are "at the money" or "out of the money", i.e. where market prices are frequently equal or even below the variable operation costs. Clearly it is then also necessary to include operation restrictions to account for the limited flexibility of real plants. Furthermore, it is evident that good price models are required in order to obtain adequate results. As pointed out in Weber (2003), different specifications may lead to substantially different results.

6.3   Two stage optimization

The two-stage optimization approach developed in section 4 is applied to the same months as the real option model in the previous section. For each day, the optimization is carried out taking into account the real prices of the day and the price expectations for the next day. Optimization results for selected weeks (Monday to Sunday) are shown in Figure 6-11 to Figure 6-13. Clearly, the higher heat demand and the higher average electricity price imply that the power plants are producing more electricity in January than in April. But peak electricity prices have also been extremely high in the July week under study, leading to an even higher peak production in July than in January. Given the lower heat demand, the electricity production of the extraction condensing turbines is not reduced by the extraction share. For example, turbine 3 produces almost 300 MW of electricity in July as compared to 230 MW in January. Whereas the coal-fired power plants are almost continuously operating during weekdays in January, their operation becomes more price dependent in the intermediate and summer season. Given the considerable start-up costs, they are usually not turned off during night time on weekdays but their output is reduced to the minimum as long as prices are low. On weekends most of them can be turned off completely if heat demand is low as shown in Figure 6-13 for the July Sunday. On Monday (first day) in the April week, a rather low production and price level is observed, too. In fact this is Easter Monday which is a public holiday in Germany. The gas-fired plants (T7, T8, T9) are mostly operated in winter and even there their operation is often limited to the day time. In the intermediate and summer season, turbine T7 is operating almost continuously in order to satisfy grid restrictions limiting heat transfer between the two district heating grids. The other units, including the oil peaking unit T9, are only run when peak electricity prices are high enough to cover also the start-up costs.

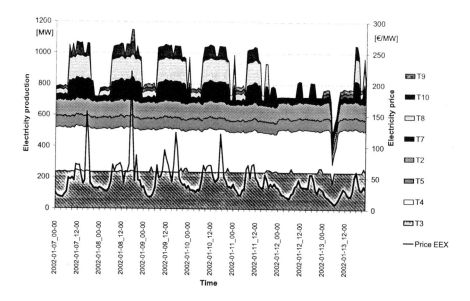

Figure 6-11. Operation of the CHP system in the week of January 7 - 13, 2002 according to the two stage optimization model

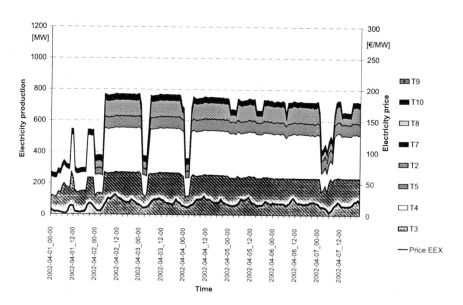

Figure 6-12. Operation of the CHP system in the week of April 7 - 13, 2002 according to the two stage optimization model

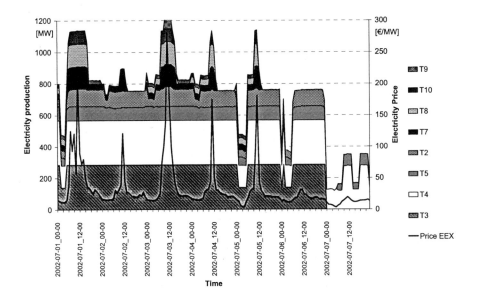

Figure 6-13. Operation of the CHP system in the week of July 1 - 7, 2002 according to the two stage optimization model

6.4 Longer Term Portfolio Management

For the longer term portfolio management, the developed approach is again applied to the cogeneration portfolio described in section 6.1.2. Thereby products with delivery in 2003 are considered. The Monte-Carlo simulations of expected spot prices in 2003 obtained from the integrated model described in chapter 4 are condensed into a lattice with five daily price scenarios.

Additionally three heat demand scenarios are generated as described in section 6.3. These scenarios are combined in order to yield a stochastic scenario lattice with 15 nodes per time step (cf. Figure 6-6). Correlations between heat demand and spot market prices are not considered since sufficient data points covering simultaneously electricity prices and heat demand are not available. Moreover, it is assumed that the expected spot market prices derived from the model are consistent with the current future prices so that the martingale property (6-66) is fulfilled[40].

---

[40] Otherwise, the valuation of the portfolio should be done under an equivalent martingale measure (cf. e.g. Ross 1977, Hull 2000, King 2002, Wallace and Fleten 2002), i.e. the simulated spot market prices should be transformed to prices consistent with the observed forward prices. However there is no unique transformation rule in the case of the

Using these scenarios, the expected electricity production quantities shown in Figure 6-14 are obtained. Obviously, electricity production is lowest during the summer months when the heat demand is low and prices also tend to be lower. Conversely, the electricity production is highest in January, March and December, the smaller value in February being mostly due to the smaller number of days in this month. The deviations between the stochastic and the deterministic model are limited, but the stochastic model leads in most months to somewhat lower electricity production quantities. This is due to the fact that the power plants are partly not operated in scenarios with low prices. Besides yielding the expected value, the stochastic optimization model provides also the opportunity to determine the range of variation for the electricity production. This is shown in Figure 6-15. The variations are larger in the summer months even though the average value is smaller since the probability that the power plants are out of the money is higher in summer than in winter.

The coal consumption (cf. Figure 6-16) is in most months somewhat lower in the stochastic than in the deterministic model. This is due to the fact that at average prices the coal fired plants are most of the time in the money. When taking into account the stochastics, it turns out that it might sometimes be more profitable to turn them off.

For the gas-fired plants, the opposite phenomenon is observed. Those are mostly out of the money for the average electricity prices but they are to a certain extent in the money in the stochastic scenarios. Consequently the gas consumption (cf. Figure 6-17) is higher in the stochastic than in the deterministic modeling. The moderate impact of the model specification on the expected electricity production is hence in fact the result of two opposite effects: the coal plants are producing less in the stochastic scenarios but the gas plants are producing more.

The value of the portfolio, corresponding to its expected profit is in the stochastic model more than 10 % resp. about 5 Mio. € higher than in the deterministic model. This is particularly a consequence of the more intensive use of gas-fired power plants in the stochastic model. As with conventional power plants, the gas-fired units with CHP will be more frequently in the money and have therefore a higher value in the stochastic than in the deterministic case. This additional profit corresponds to the option value of the power plants in the given system.

What risk is associated with this power plant portfolio depends on the hedges which have been undertaken. This will be analyzed in more detail in

---

electricity markets, given that a bundle of spot market products (hours) corresponds to one forward market product (peak or base monthly contract), cf. also chapter 7 sections 4 and 6.

chapter 7. But hedging with the expected electricity production and the expected gas and coal consumption certainly contributes to reduce considerably the price risk with respect to forward price changes (cf. chapter 7, section 6). Yet the operator still carries the risk that the prices in the single hours develop differently than the price of the standard products. If the production profile then does not correspond to the load shape of the standard products base and peak, a basis or delivery risk remains with the operator (cf. chapter 7). On the other hand, the operator may realize additional profits by dynamically hedging the portfolio as soon as price expectations for the spot market change. This allows realizing the option value of the power plants.

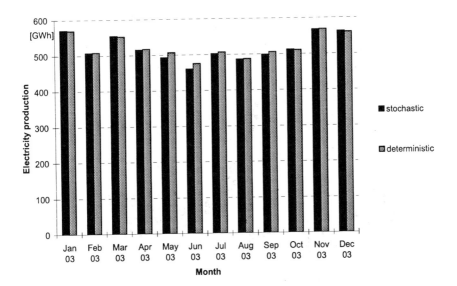

Figure 6-14. Expected average electricity production in the year 2003 for the stochastic and the deterministic model approach

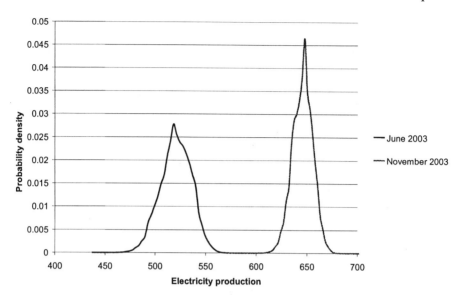

Figure 6-15. Variation of electricity production in the year 2003 for selected months in the stochastic model

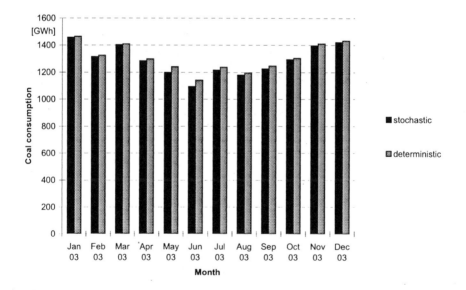

Figure 6-16. Expected average coal consumption in the year 2003 for the stochastic and the deterministic model approach

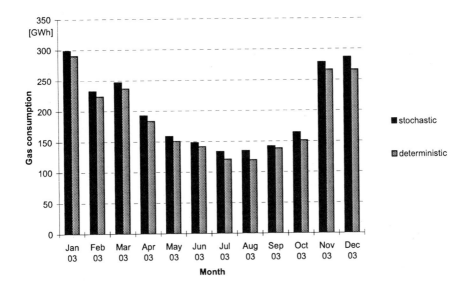

Figure 6-17. Expected average gas consumption in the year 2003 for the stochastic and the deterministic model approach

Chapter 7

RISK MANAGEMENT AND RISK CONTROLLING

Given the large uncertainties associated especially with energy prices, the methods for operational planning discussed in the previous section have to be complemented by methods for risk management. In the finance sector, risk management is clearly recognized as a key company activity and required by law (e.g. KonTraG[41] in Germany) and the supervising institutions (e.g. BAFIN[42] in Germany). Risk management has as major objective to prevent major damages which might endanger the future of a firm. Therefore a risk management system is required which has to be anchored at the top management level through a risk board and which puts into practice a general risk policy defined at the top management level (cf. Dörner et al. 2000, Rudolph, Johanning 2000). This requires implementation both in the functional and the process organization of the firm. In a process oriented view, risk management notably comprises the steps of (cf. Horváth, Gleich 2000, Walther 2000):

- Risk identification
- Risk measurement and quantification
- Risk aggregation
- Risk assessment
- Risk steering and control
- Risk monitoring.

---

[41] „Gesetz zur Kontrolle und Transparenz im Unternehmensbereich" (N.N. 1998) = Law for control and transparency in companies

[42] „Bundesaufsichtsamt für Finanzdienstleistungen" = Federal supervisory agency for financial services

These processes form a feedback loop which should ensure the quality of risk management. They have to be under constant supervision of the risk board.

In the energy field, risk management has also been a topic for many years (cf. e.g. Fusaro 1998, Bergschneider et al. 1999, Pereira et al. 2000, Voß, Kramer 2000 and the literature overview in Dahlgren et al. 2003) but still the tools for risk management are less standardized and well established as in the finance field. On the one hand, many activity fields of energy companies are less exposed to daily market risk (e.g. in the retail field), on the other hand the technical risks and the non-applicability of standard price models make consistent risk management an even larger challenge.

Therefore first the relevant risks in energy (and especially electricity generation and trading) companies are classified in the following (section 1). Then different concepts for aggregate risk measures are discussed (section 2), since these form the key channel for measuring, communicating and controlling quantitative risks. In particular the new concept of Integral Earnings at Risk (IEaR) is introduced and in section 3 put into relationship to the required risk management strategies in incomplete markets. Section 4 then presents models for quantifying the identified price risks in electricity markets, and section 5 describes possibilities for modeling the risks associated with physical assets, notably power plants. These approaches are then applied in section 6 to the portfolio discussed in the previous section.

1. TYPOLOGY OF RISKS

Besides market risks also other external and internal risks have to be considered within in a systematic risk management approach (cf. also Pilipovic 1998, Bergschneider et al. 1999, Hufendiek 2002). These broad categories may be further specified considering the following aspects.

1.1 Market risks

These comprise notably:
- Price risks,
- Quantity risks,
- (Market) liquidity risks.

Price risks: They include besides the risks on the electricity market also the risks of fuel price changes and risks of price changes for other input factors. But given the high volatility of electricity spot market prices, these risks need to be scrutinized most carefully. Furthermore the price risks on the future and forward markets have to be looked at, given that here often huge quantities are traded. Last but not least interest rate changes and changes in currency exchange rates are price risks, which may be of considerable importance for the company – e.g. if the primary energy is billed in dollars, but the electricity delivered in another currency such as Euro.

Quantity risks: These are of particular importance in the electricity market, given the non-storability of electricity. On the supply side, power plant outages have notably to be considered, whereas on the demand side load fluctuations among customers with full service contracts are most important. Moreover the positive correlation with price risks has to be taken into account.

(Market) liquidity risks: Given that electricity is produced in power plants with largely given capacities (in the short term), the liquidity in the market is limited. Limited capacities will of course translate in rising prices, but it is worth considering this as a separate risk factor, since possibly no market equilibrium is attained, even at extreme prices (cf. the events at the German power trade EEX on 17 December 2001 and 7 January 2003). Limited liquidity may also create problems when open positions have to be closed in derivative markets.

1.2   Other external risks

Counter-party risks: The Enron collapse and the subsequent liquidity reduction in North American and European energy trading have clearly demonstrated the necessity to pay attention also to this type of market related risks. The power exchanges offer here a clear advantage by taking over this risk for the trading parties.

Political risks: As already mentioned in chapter 2, political uncertainty may considerably affect the longer term market position of energy companies. They have therefore to be considered, especially when it comes to strategic risk management.

Financial risks: These include particularly the risks related to financial liquidity, but also the risks a company is bearing through the financial participations it is holding.

1.3   Internal risks

Process and project risks: Setting up new processes or projects bears always some risk. These have to be addressed clearly already at the planning stage and have to be continuously monitored in the following.

Personal risks: It is well known that humans make errors and this can constitute a risk for a company. But there are also other risks related to staff members: they may get ill or they may quit the company. In this case, essential knowledge may be lost for the enterprise and this might even endanger the continuation of the business.

IT-risks: The electricity generation, trading and retailing business is extremely dependent on a well-functioning information technology. Therefore any risk related to the functioning of hardware and software equipment has to be taken very seriously.

Model risks: Last but not least also the possibility has to be considered that the models used for measuring risk and valuing positions are themselves imperfect or faulty. Therefore the models used have to be tested and evaluated regularly.

2.   AGGREGATE RISK MEASURES

Of course not all risks mentioned in the previous section can be quantified easily. Some are rather important, but still may be better addressed verbally than through some statistics. Yet in those fields, where quantification is possible, it allows to provide an objective picture of the risk situation of the company. For a description of the overall situation and an adequate support of managerial decisions it is thereby necessary to describe the risk situation through aggregate measures. Currently the most popular measure of that kind certainly is the value at risk (VaR) concept. This will be briefly reviewed in section 2.1. In section 2.2, alternative proposals for describing the market risk situation of a company are discussed and in section 2.3 alternative statistical measures for expressing the risk are reviewed. Finally, section 2.4 is devoted to the description of a new measure for assessing the total market risk.

2.1   Value at risk

The concept of value at risk (VaR) has been introduced in the nineteen-nineties in the finance sector (cf. notably Group of thirty 1993, J.P.Morgan 1994). It aims at assessing the whole risk of a portfolio through one number

R_{VAR}. It provides the following statement: We are α percent certain, that we will not lose more than R_{VAR} Euro in the next d days (cf. Hull 2000, p. 342). The *VaR* is thus a function of the holding duration d (usually between 1 and 30 days) and the confidence level α (99 % or 95 %). Analytically, this may be expressed as:

$$R_{VAR}(V,d,\alpha) = \left| F(V(\Pi,t_0 + d); 1 - \alpha) - V(\Pi,t_0) \right| \tag{7-1}$$

Hence the value at risk corresponds to the difference in absolute terms between the quantile $1 - \alpha$ of the cumulative distribution function $F(.)$ of the value V of portfolio Π at time $t_0 + d$ and the current value $V(\Pi,t_0)$. Thereby the martingale property is usually assumed for the value of V, i.e. $V(\Pi,t_0) = E[V(\Pi,t)]$.

The value of the portfolio is a stochastic variable, which may be determined by adding up the single positions y_k valued at their current prices $p_{k,t}$:

$$V(\Pi,t) = \sum_{k \in \Pi} p_{k,t} y_{k,t} \tag{7-2}$$

Value changes correspondingly may be written:

$$\Delta V(\Pi) = V(\Pi,t) - V(\Pi,t_0) = \sum_{k \in \Pi} \left(p_{k,t} y_{k,t} - p_{k,t0} y_{k,t0} \right)$$
$$\approx \sum_{k \in \Pi} \left(\Delta p_k y_{k,t0} + p_{k,t0} \Delta y_k (\Delta p_k, \Delta t) \right) \tag{7-3}$$

They are hence composed of the effects of price changes and the effect of quantity changes. Second order effects due to the interaction between the price and the quantity effect are accounted for in the first formulation on the right hand side, but they are neglected in the second. Among the quantity changes only those are considered, which are a consequence of the price changes, such as the exercise of an option, and those which are directly linked to the time variable, e.g. load fluctuations. By contrast, deliberate changes in positions through sales or purchases are not considered when computing the *VaR* since it aims at expressing the risks associated with the *current* portfolio.

This necessitates the analysis of possible price paths (cf. section 4) and the analysis of related or unrelated quantity changes (cf. section 5). For the computation of the VaR, mostly two methods are applied: either the Value at risk is calculated analytically or it is determined through Monte Carlo simulations[43]. The analytical calculus for the VaR is however only possible under restrictive assumptions, concerning notably the distribution of price changes and the valuation of options. Therefore, we will focus in the following on Monte Carlo approaches to VaR computation. On beforehand, several other concepts for assessing the market risk will be scrutinized.

2.2 Alternative concepts for market risk

The strength of the VaR concept clearly is that it provides a unique, rather easily understandable number to quantify the risk exposure of a company. It has therefore found widespread application in the banking sector, yet its applicability to producing companies such as electricity generators has been repeatedly questioned (cf. Henney, Keers 1998, Dahlgren et al. 2003, Denton et al. 2003). Also for the banking sector, alternative risk measures are discussed (e.g. Johanning, Rudolph 2000). The key criticisms become clear when a look at proposed alternative measures is taken:

Earnings at risk (Dorris and Dunn 2001): This risk measure only considers the contracts (physical or financial) which come to delivery within a prespecified time period, e.g. the next accounting year. This can be attractive in the financial sector to assess the risks for profits or losses in a given accounting period. In the energy sector, it similarly emphasizes on the upcoming results in the period under consideration instead of analyzing the changes in market value, which correspond to the sum of all discounted future cash flows (at least under a certain value model, cf. Rudolph 2002).

Cash flow at risk (Guth and Sepetys 2001): For Earnings at risk, the date of delivery of contracts is the criterion for inclusion or exclusion in the risk calculation. In the Cash flow at risk approach, additionally the date of settlement, which may be the day of delivery or up to thirty days later, is considered so that this risk measure directly targets the financial liquidity of the company.

Profit at risk (Henney, Keers 1998, Barnwell 2001): Again this risk measure is rather similar to Earnings at risk, considering the revenues from the current portfolio during a pre-specified period (one or several years). But the reasoning behind it is different: the emphasis is laid on the fact that for

---

[43] Sometimes the use of historical price simulations is mentioned as a third alternative (e.g. Hull 2000, p. 356). However, if correlations between price changes shall be accounted for adequately, this requires longer consistent price series than available for most electricity market products.

many assets in the electricity industry, no exact market based hedge can be found, which would allow closing a position at the latest at the end of the pre-specified duration (the holding period). For example, almost no hedges for load demand for single hours may be found on the derivative market, thus risk has to be carried on until delivery of the electricity. At the same time, price risks are much higher at the moment of delivery (spot market) then in the next days, when a product is traded on the forward market.

2.3   Alternative statistical risk measures

Besides the question of which earnings, values or cash flows should be included when measuring risk, the statistical criterion chosen to describe risk may also be questioned. Before the appearance of value at risk, the *variance* or *standard deviation* of the distribution have traditionally been used to describe risk, notably within the portfolio selection model of Markowitz (1952) and the capital asset pricing model of Sharpe (1964) and Lintner (1965) (cf. e.g. also Ross 1977, Varian 1992). The use of the variance as a risk measure is justified if either the distribution of value changes follows a normal distribution or if the preferences of the decision maker correspond to a quadratic utility function (cf. Varian 1992, Rudolph 2002). But instead of choosing an arbitrary and not directly interpretable weighting factor for the variance in the quadratic utility function, the use of the value at risk concept leaves the decision maker(s) with the choice of a confidence level, which can easily be interpreted as an acceptability threshold.

However this threshold characteristic of the Value at risk is also problematic, both mathematically and for interpretation. Mathematically, the inconvenience of Value at risk defined as a quantile is that it might be non-continuous and furthermore that it does not form a coherent risk measure as defined by Artzner et al. (1999). From an interpretational point of view, it seems questionable that the height of the losses below the threshold considered does not influence at all the Value at risk. If losses do not exceed 1 Mio. € with 99 % certainty, then it makes no difference whether losses exceed 10 Mio. € with 0.5 % probability or whether they remain limited to 1.5 Mio. € in this more extreme case.

As an alternative, the *conditional value at risk* (*cVaR*), also sometimes called *tail-VaR* or *mean excess loss,* is therefore proposed. For any stochastic variable *V* with finite first moment, it is defined as the probability weighted sum of all deviations from the mean below a certain probability threshold 1 - α (cf. Rockafellar and Uryasev 2000, Uryasev 2000):

$$cVaR(V,\alpha) = \left| E[V|V \le F(V;1-\alpha)] - E[V] \right| \tag{7-4}$$

This measure clearly has an integral character and accounts also for really extreme events below the threshold. Since it is smooth and convex, it is also particularly suited for stochastic optimization (cf. also Römisch 2001, Schultz and Tiedemann 2002).

2.4   Integral earnings at risk

The preceding discussion has highlighted three major shortcomings of the traditional VaR concept when applied to the electricity industry:

- It assumes that market prices for all assets are available and can be used for a mark-to-market valuation[44].
- It assumes that open positions may be closed during the holding duration through market transactions.
- It does not account for the further downside risk which still exists beyond the chosen threshold.

To remedy to these flaws and still obtain a workable solution, which serves the overall purpose of controlling risks which might endanger the future existence of the company, a new risk measure is proposed here: *Integral earnings at risk*. It aims at measuring the impact of day-to-day market fluctuations on the accounts of a company. It includes:

- the conditional earnings at risk of all tradable assets,
- the conditional earnings at risk of all non-tradable assets, adjusted for hedges with standard market products after the holding duration,
- a measure of market liquidity risk if end-user load has been contracted.

One may first note that given the limited lifetime of energy market assets like forwards and futures, the first part of earnings at risk is the same as value at risk applied to an energy trading portfolio, except that the integral conditional risk measure is used instead of just the quantile. For the second part, this corresponds to a profit at risk approach, except that hedging with standard market contracts is allowed to (partly) close positions. It could therefore also be labeled "partial profit at risk".

The rationale for this choice becomes evident when looking at the major positions in the accounts of a vertically integrated company, comprising generation, trading and retail businesses[45].

---

[44] For non-standard assets also in finance a "mark-to-model" approach is proposed as a workable alternative (cf. Rudolph 2002). The assets are then valued using standard market prices and some (modelling) assumptions how the value of the asset is linked to the observable market prices.

[45] Obviously, the use of integral earnings at risk is not limited to fully integrated companies. E. g. for a pure generation company, the measure may be applied to the generation assets and all sales contracts on the wholesale market concluded by the company.

In the *trading business*, the majority of all open positions are clearly tradable assets, purchased or sold on financial or physical wholesale markets. These assets may be valued mark-to-market and open positions may be closed within a limited time period known as holding duration. Therefore the earnings at risk correspond to the value at risk for this part of the portfolio. For non-standard products, the risk measure should also comprise risks occurring after the end of the holding duration, given that it will be difficult to close these positions during the limited holding duration. Yet the use of earnings at risk as a measure for this risk component will be sufficient[46], since these trading positions usually only generate earnings over a limited delivery period. The period covered by the earnings at risk should thereby obviously comprise the delivery period of the non-standard product. Furthermore the earnings at risk of the non-standard product (or rather the portfolio including non-standard products) should be adjusted for hedging opportunities offered by the market, because otherwise the risk is overstated.

The main assets in the *generation business* are obviously the power plants owned. These may be complemented by long-term purchase contracts or participations in shared generation assets. A valuation of power plants using a mark-to-market approach is not directly feasible given that a liquid market for power plants with daily price notations does not exist. So on a day-to-day basis, only the revenues generated by this power plant and the fuel costs incurred may be valued. And since long-term forward notations are not available, earnings at risk with a limited time horizon is obviously the only practicable alternative. Again, these risks should be considered just for those parts of the portfolio which can not be closed when using standard market products. For the "standardisable" part, the (conditional) risk during the holding period (corresponding to traditional value at risk) is to be considered only.

The *retail business* mostly holds assets in form of full service contracts with electricity end users. These may be balanced by standard or non-standard forward and/or future contracts to cover the expected demand. As with generation assets, a closing of positions with standard products can be assumed after the holding duration and consequently positions have to be split up in a standard and a non-standard part. The standard part can be valued using traditional (conditional) value at risk measures whereas for the remainder earnings at risk during delivery have to be computed. Quantity risks are an important risk in the retail business and have therefore to be

---

[46] The term of profit at risk has been registered as a trade mark by KWI (cf. e.g. KWI 2003) and strongly promoted in the electricity industry. But to the author's understanding, it corresponds to the concept of earnings at risk with a clear emphasis on the fact that quantity risk is a key component of energy market risk.

included in the earnings at risk calculation, especially given that they may not be compensated via the standard products available through trading. Additionally, the risk of lacking market liquidity to fulfill the physical delivery obligations should be taken into account.

A direct valuation of the market liquidity risk using market prices is difficult. The prices for regulatory power could be used as a proxy, however this hinges on two assumptions: Firstly, there must be sufficient reserve capacity available to physically maintain the balance of supply and demand even if the day ahead market does not establish a market clearing price, and secondly, the price used for regulatory power has to reflect the costs of this risk to the company[47]. Another alternative is to recur to the "value of lost load" approach (cf. e.g. Telson 1975, Wacker and Billinton 1989, Kariuki and Allan 1996, Willis and Garrod 1997) which obviously raises the question what value should be used there. It can be possibly derived from the value of load shedding which is implicitly included in full service contracts with demand flexibility, if those exist. Another possibility is to recur to customer surveys, which however are subject to various potential biases. If none of these informations are available, it has to be set arbitrarily, if possible in accordance with the regulation authority.

3. INTEGRAL EARNINGS AT RISK AND RISK MANAGEMENT STRATEGIES IN INCOMPLETE ELECTRICITY MARKETS

The integral earnings at risk (IEaR) measure has been introduced above as a means to remedy to some of the flaws perceived in the construction of conventional value at risk. In the following it is shown that the integral earnings at risk indeed is suitable for risk management purposes, by demonstrating that it prompts for immediate action when the market structure requires it, without imposing unnecessarily strong constraints on business operation or hastening prematurely the action.

Let us therefore look at the risks associated with the holding of a complex asset and the possible risk management strategies. By complex asset any non-standard product is meant, which is not directly traded on the wholesale market. But more specifically we will look at power plants or

---

[47] If the so determined marginal costs to the company are below the marginal revenues required to install new generation capacity, long-term supply adequacy will be a system or societal problem which might necessitate a change in market design (cf. de Vries, Hakvoort 2003 and chapter 9).

other assets with optionalities, since these constitute the largest challenge for risk management.

The considerations are limited to a predefined delivery period, which should correspond to the time horizon covered by the trading business (around two years). Beyond this time horizon, hardly any reliable market quotes are available which may be used for risk quantification. So the longer the time horizon is, the more the situation corresponds to one of decision making under true uncertainty instead of risk (cf. section 3.3.1). This has of course to be considered within strategic risk management, but is beyond the realm of operational risk controlling measures.

What can happen to the value of the power plant – or equivalently to the earnings it will generate - between today and the period of delivery (cf. Figure 7-1)?

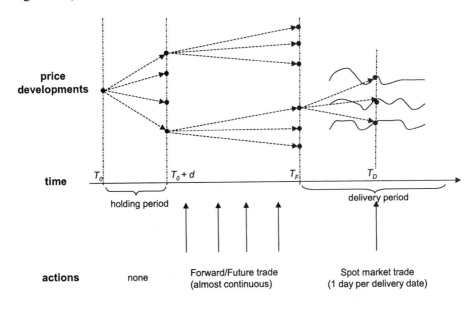

Figure 7-1. Price developments and recourse actions for risk management

- The forward/futures prices may change between today and the end of the forward market
- The average spot market prices may deviate from the last recorded forward/future prices (cf. chapter 4).
- The spot market may get illiquid.
- The power plant option may not be available due to outages or other non-availabilities.

What can be done to mitigate or reduce these risks? First, one should note that the risk of spot market illiquidity is not of major relevance for power plant holders, on the contrary this offers opportunities for excess profits. It is therefore not considered further here. But to mitigate the other risks, the following actions are conceivable:

- Standard products may be sold (or purchased) at the forward/future markets to fix some of the revenues
- Non-standard products may be sold (or purchased) on the derivative markets
- The electricity produced may be sold at the spot market or contractual obligations may be fulfilled through electricity purchases at the spot market.
- The physical availability of the power plant may be increased through technical measures.

For the second and the fourth action, the feasibility is rather uncertain, therefore risk management should not rely on them. But the purchases and sales at the forward/future market and the spot market are almost certainly available as recourse actions and should therefore be taken into account in a standardized risk measure. Hence the remaining risks may be divided into two large groups, as depicted in Figure 7-2, the *short-term risk* before open positions can be closed in the futures and forward market and the *delivery risk* after the end of the forward and future market. For a quantification of these risks, the course of action has to be looked at in reverse direction in order to identify what risks can be avoided at later stages and which have already to be taken care of right now.

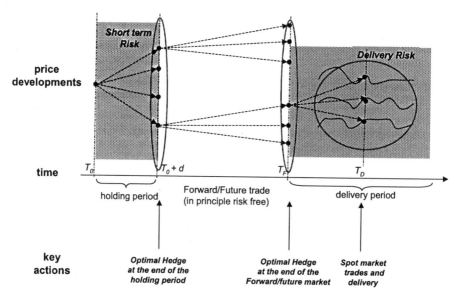

Figure 7-2. Risks and key actions in incomplete electricity markets

At the time of delivery t_D, the value V_{PL} of the earnings of a power plant with rated output P_{PL}^{EL} corresponds to the value of the related option pay-off function (cf. chapter 6, section 3), corrected for quantity risks. If we take the model of the spark-spread option as reference, neglecting in a first step the quantity risks, this value at exercise will correspond to the difference between electricity price $p_{t_D,S}^{EL}$ and fuel costs $p_{Strike} = h_{PL} p_{t_D,S}^{FU}$, if this difference is positive, and zero else:

$$V_{PL}\left(t_D, \mathbf{p}_{t_D,S}\right) = \max\left(0, p_{t_D,S}^{EL} - h_{PL} p_{t_D,S}^{FU}\right) P_{PL}^{EL} \qquad (7\text{-}1)$$

This holds independently of the decisions taken on the forward market. According to the mark-to-market principle, forward contracts will be valued at spot market prices and missing or excess quantities will be traded on the spot market.

Going one step back to the time of closure of the forward market, the value of the power plant will be a stochastic variable, depending on the stochastic price vector $\mathbf{p}_S = \left\{p_{t_D,S}^{EL}\right\} \cup \left\{p_{t_D,S}^{FU}\right\}$ at delivery, which in turn is a function of the forward price at market closure:

$$V_{PL}\left(T_F, \mathbf{p}_S \middle| p_{T_F}^{FW}\right) = \sum_{t_D \in T_D} \max\left(0, p_{t_D,S}^{EL} - h_{PL} p_{t_D,S}^{FU}\right) P_{PL}^{EL} \qquad (7\text{-}2)$$

This value comprises a summation over all times of delivery t_D included in the delivery period T_D for which the forward/future product is valid. It sets implicitly the quantity sold at the spot market y^{Spot} equal to the power plant output $O_{PL} P_{t_D,S}^{EL}$. This physical balance has to be modified somewhat if a quantity y^{FW} has been contracted already on the forward market:

$$
\begin{aligned}
O_{PL,t_D,S} P_{PL}^{EL} &= y_{t_D,S}^{Spot} + y^{FW} \\
\Leftrightarrow \qquad y_{t_D,S}^{Spot} &= O_{PL,t_D,S} P_{PL}^{EL} - y^{FW}
\end{aligned}
\qquad (7\text{-}3)
$$

The value of the total portfolio is then:

$$V_{tot}\left(T_F, \mathbf{p}_S \middle| p_{T_F}^{FW}\right) = \sum_{t_D \in T_D}\left(-p_{Strike}O_{PL,t_D,S}P_{PL}^{EL} + p_{T_F}^{FW}y^{FW} + p_{t_D,S}^{El}y_{t_D,S}^{Spot}\right)$$

$$= \sum_{t_D \in T_D}\left(\max\left(0, p_{t_D,S}^{EL} - p_{Strike}\right)P_{PL}^{EL} + \left(p_{T_F}^{FW} - p_{t_D,S}^{EL}\right)y^{FW}\right)$$

$$(7-4)$$

In Figure 7-3 to Figure 7-5, this value is depicted for a single time segment t_D as a function of the spot market price with the forward/future quantity contracted set to 0, $P_{PL}^{EL}/2$ or P_{PL}^{EL}. The expected value is independent of the quantity of electricity y^{FW} contracted on the forward market, if the martingale property $p_{T_F}^{FW} = E\left[p_{t_D,S}^{EL}\right]$ is satisfied. But the maximum downside risk depends on the choice of y^{FW}. As long as y^{FW} is in the interval $[0\ P_{PL}^{EL}]$, the minimum portfolio value occurs in a neighborhood of the strike price. And which forward contracting quantity y^{FW} yields the lowest risk, depends only on the relationship between $p_{T_F}^{FW}$ and p_{Strike}.

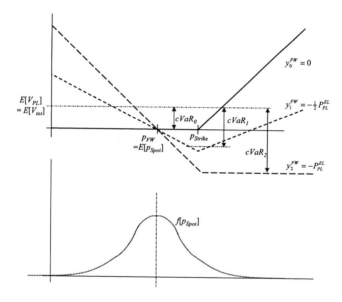

Figure 7-3. Portfolio value and conditional earnings at risk for a portfolio with power plant, forward sales and spot sales as a function of spot price and forward quantity (power plant out of the money)

If the forward price is below the strike price (the option is out of the money), the risk minimizing decision is to sell no forwards at all (cf. Figure

7-3). On the contrary, if the option is in the money at $p_{T_F}^{FW}$, the optimal strategy is to sell the whole power P_{PL}^{EL}, even if it is far from certain that the option is exercised (cf. Figure 7-4). Only in a narrow band around the strike price p_{Strike}, a mixed strategy may lead to a slight reduction in the conventional and the conditional value at risk as indicated in Figure 7-5.

For the case with multiple time segments t_D, no such simple rule can be determined. But one may observe that the minimum risk strategy is to sell everything or nothing if the errors in the time segments are uncorrelated, because the minimum portfolio value in each time segment arises for $p_{t_D,S}^{EL} = p_{Strike}$ and for uncorrelated errors, the total cVaR is (approximately) determined by adding up the single minima. If the sum of the minima is positive for one value of y^{FW}, it will be positive for any value of y^{FW} and vice versa. Thus the maximum of the minima is either obtained for $y^{FW} = 0$ or for $y^{FW} = P_{PL}^{EL}$.

If on the contrary, all errors are perfectly correlated, the risk optimal strategy is to sell a proportion equal to the share of time segments when the power plant is in the money. This can be seen from Figure 7-6.

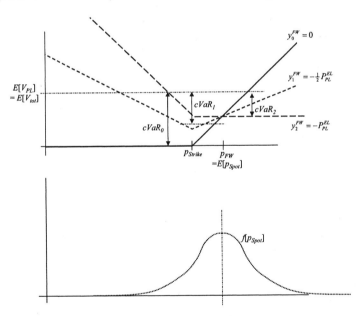

Figure 7-4. Portfolio value and conditional earnings at risk for a portfolio with power plant, forward sales and spot sales as a function of spot price and forward quantity (power plant in the money)

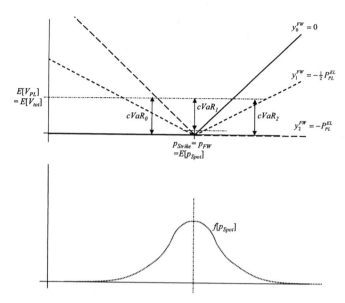

Figure 7-5. Portfolio value and conditional earnings at risk for a portfolio with power plant, forward sales and spot sales as a function of spot price and forward quantity (power plant at the money)

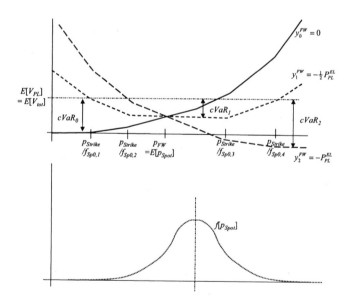

Figure 7-6. Portfolio value and conditional earnings at risk for a portfolio with several exercise time points and different price levels but correlated errors

In this case, the power plants and associated revenues may be ordered according to an increasing ratio of strike price to expected price in hour t_D, and a value function similar to those in Figure 7-3 to Figure 7-5 can be drawn for the overall portfolio as a function of the single stochastic factor. As shown in the figure, the conditional value at risk is lowest if the value function is horizontal when crossing the forward price. This corresponds to selling as much on the forward market as there are options in the money.

For more complex real options, it may be even more difficult to give the optimal decision rule for the minimum-risk portfolio in time T_F. In fact, the optimal hedge can only be obtained as the result from a stochastic optimization problem, similar to the ones discussed in chapter 6. This optimal hedge may differ from the hedge obtained from a deterministic production planning, much as a difference between deterministic and stochastic production scheduling was observed in chapter 6.

Furthermore the optimal hedge at the closure of the forward market obviously depends on the forward price level. Viewed from today, the earnings at risk for the delivery period are hence not uniquely determined by the current forward prices, but depend on the possible price developments between time T_0 and T_F.

A further complication arises from the fact that the minimum-risk portfolio at T_F is not risk free in the financial, continuous hedging world. With continuous hedging, one aims at constructing a risk-free portfolio by combining options and forwards in a way that the resulting portfolio is risk-free (or at least risk-minimal). If price shocks on forward prices have similar effects on all spot market scenarios, the quantity of forwards required for a risk-free portfolio corresponds to the average probability of exercising the options at the exercise time. This is the "expected operation hedge" or "naïve delta hedge" mentioned in chapter 6, section 5[48]. But clearly an all-or-nothing hedge as observed in the simple examples of Figure 7-3 to Figure 7-5 is not equivalent to such a probability based hedge, and also the other hedge patterns discussed above do not necessarily fulfill the property of being risk-free in a setting with continuous hedging. A rehedging would therefore be needed shortly before the closure of the forward market. Expressed differently: As a consequence of the change in market operation

---

[48] As shown below, an additive model linking forward prices and hourly spot market prices must hold if the expected operation hedge is to be a delta hedge. Therefore it is labelled "naïve" here.

mode, the optimal hedging changes, too[49]. This rehedging can in principle be done at zero cost and zero instantaneous risk – at least with liquid markets.

Given these difficulties to determine and manage exactly the optimal hedge at the end of the forward market, and furthermore the necessity not to be overoptimistic in risk management, it seems more appropriate to take a "reasonable hedge" at the end of the forward market as basis for determining the delivery risk, instead of assuming that the optimal hedge will be implemented. A candidate for such a reasonable hedge is certainly the "expected operation hedge", determined either on the basis of the expected hourly spot prices or on the basis of the stochastic spot prices. Which alternative is preferable will be investigated empirically in the following.

In the (almost) continuously operating forward and future market, a delta hedge may in theory be adjusted continuously at zero cost to remain risk free (cf. e.g. Hull 2000). Consequently, the optimal hedge at the end of the holding period is again a delta hedge, corresponding (under certain assumptions) to the then perceived average exercise probabilities of the power plant option. The second part of risk, which remains in any case to the utility, is thus the risk of losses during the holding period. This is the short-term risk mentioned in Figure 7-2.

One may argue that also during the forward trading period no continuous adjustment is possible, but instead some adjustment time, corresponding e.g. to the holding period, should be taken into account for each adjustment. Yet one major reason for introducing a non-negligible holding period is that the forward market may not be liquid enough to allow hedging a complete portfolio within very short time notice. After having established a hedged position, the quantities to be dynamically rehedged are on the contrary much smaller. So they should be easier to place on the market and also the risks associated with them (if there are any) are much lower and may be neglected in a first step.

4.   PRICE MODELS FOR RISK CONTROLLING

For determining the integral earnings at risk, it is thus necessary to have workable price models for capturing both the short-term risk and the delivery risk. One possibility is to develop consistent models linking the evolution of forward prices directly to the spot prices (e.g. Schwartz 1997,

---

[49] This may be part of the explanation, why in the continental European markets high trading volume and corresponding volatility are especially observed just before the closure of the forward market.

Hilliard and Reis 1998, Karesen and Husby 2002, Lucia and Schwartz 2002). However these approaches rely on no-arbitrage arguments not only as far as future spot and the corresponding forward prices are concerned (cf. equation (6-67)), but also for the relation between current spot and forward prices. Such a no-arbitrage argument is clearly only applicable for storable commodities or if in some other way possibilities of storage are provided, such as hydro storage in the Nordic electricity market. Otherwise, markets are obviously incomplete, i.e. there are more risk factors than products available to hedge the risk.

Harrison and Kreps (1979) have shown that in complete markets the absence of arbitrage is equivalent to the existence of a unique so-called equivalent martingale measure $\tilde{\Phi}$, i.e. a probability measure (or loosely speaking: a set of modified probabilities) such that equation (6-67) is modified to:

$$p_{FW,T_0,T}^{EL} = E_{\tilde{\Phi}}\left[\overline{p}_{Spot,T}^{EL}\right] \qquad (7\text{-}5)$$

where $E_{\tilde{\Phi}}[\ ]$ is the expectation operator under the modified probability measure $\tilde{\Phi}$. Under this modified probability, a risk-neutral valuation of assets can be performed and the results are then also valid with the original "physical" probability measure.

If markets are incomplete, in general multiple equivalent martingale measures exist, i.e. the value of an asset (and the associated risk) is not uniquely determined (cf. e.g. Carr et al. 2001, Dias 2003). This theoretical ambiguity translates also into a practical ambiguity, since for a given forward price $p_{FW,T_0,T}^{EL}$ of a monthly product there is no unique rule to determine the corresponding expected hourly spot prices under the equivalent martingale measure. One natural specification is an additive superposition of forward price $p_{FW,T_0,T}^{EL}$ and hourly price differentials $\Delta p_{Spot,t_D}^{EL}$:

$$p_{Spot,t_D}^{EL} = p_{FW,T_0,T}^{EL} + \Delta p_{Spot,t_D}^{EL} \qquad (7\text{-}6)$$

with the average of $\Delta p_{Spot,t_D}^{EL}$ over all time segments t_D within the delivery period T being zero.

Another obvious choice is a multiplicative one:

$$p_{Spot,t_D}^{EL} = p_{FW,T_0,T}^{EL} k_{Spot,t_D}^{EL} \qquad (7\text{-}7)$$

where the average of the hourly price factors k_{Spot,t_D}^{EL} should equal one. If the average is different from one, the obtained price distribution for p_{Spot,t_D}^{EL} can not be an equivalent martingale measure, because the value of a forward contract would then be different using current forward prices and using the equivalent spot prices. So any price process to be used for an arbitrage pricing based valuation should yield average spot prices consistent with current forward price quotations.

Obviously the integral modeling approach developed in chapter 4 provides a good starting point for modeling the spot market prices and associated risks also over a period of several years, allowing to cope hence with the delivery risk identified in the previous section. However the prices have to be adjusted to the observed forward quotes to obtain a consistent basis for valuation and risk assessment. Here a multiplicative correction term is introduced analogously to equation (6-66). The multiplicative correction seems more appropriate since it implies that forward price changes will have proportional effects on all spot prices, affecting hence higher prices in absolute terms more than lower ones.

Besides the variations in electricity spot prices also risks of forward price variations have to be considered to cope with the short-term part of the integral-earnings-at-risk concept. Here an additional approach is needed, since the integral market simulation model accounts for variations in fuel prices but not in forward electricity prices[50]. Thereby it is particularly important to capture adequately the dependency between the risks for different products through multivariate modeling.

In traditional VaR models in the finance world, geometric Brownian motion is assumed for stock and forward prices (cf. chapter 4, section 1) and thus the logical extension to the multivariate case is the assumption of a multivariate normal distribution for the relative price changes of multiple products. As discussed in chapter 4, the assumption of normally distributed price changes is however rather problematic in the case of electricity prices. For spot prices, mean reversion is a key point to be taken into account. For forward prices, mean reversion is less an issue but the existence of "fat tails", i.e. higher probabilities for extreme events than under a standard

---

[50] Those can however be interpreted in the context of the integral modeling approach as changes in the expectations on available generation capacities. According to the fundamental model part, such changes should translate into variations of spot prices in a non-linear, complex way.

normal assumption, is a major concern. In principle, extreme value theory (cf. Smith 1990, Embrechts et al. 1997) offers a direct approach for analyzing the frequency and distribution of extreme events. However, extreme value theory for multivariate time series is still at an early stage (cf. the discussion in Diebold et al. 1998, Embrechts 2000) and moreover the application of extreme value theory to time series with autocorrelation and/or volatility clustering is also at its infancy (cf. Diebold et al. 1998). Therefore, a closer look at other approaches for coping with non-normally distributed multivariate errors seems more appropriate. In particular we will look in the following at the multivariate generalization of the GARCH models (cf. chapter 4, section 2.3.2) and at possibilities for deriving multivariate models with regime switching (cf. chapter 4, section 2.3.4).

4.1 Multivariate GARCH models

In principle multivariate GARCH models are suited for the analysis of the simultaneous price changes of multiple products. As discussed by Engle and Mezrich (1996), various model specifications are possible ranging from simple scalar GARCH, which just involves a constant correlation matrix and two GARCH parameters, up to the BEKK representation of vector GARCH models (Engle and Kroner 1995) or multivariate component models (Engle and Lee 1993), which involve the estimation of two or more full matrices instead of each of the scalars in univariate GARCH models. Clearly, these complex models are very difficult to handle and estimate if multiple electricity products, such as base and peak products for single months, quarters and years, are to be considered. The time series for monthly products often comprise less than 200 observations, making a parameter estimation for more than 20 parameters a rather delicate exercise. On the other hand, the simple scalar GARCH seems too restricted in flexibility to cope with the complex and time-varying properties of electricity market products. Therefore the constant conditional GARCH model (cf. Bollerslev 1990) has been chosen as a model of medium complexity for assessing interdependencies between products with simultaneous correlation patterns[51]. It is described through the equation:

$$\sigma_{k,l,t} = \rho_{k,l}\sigma_{k,t}\sigma_{l,t} \qquad (7\text{-}8)$$

---

[51] Another possible candidate would be the factor ARCH models introduced in Engle et al. (1990). Yet these require first a factor decomposition before GARCH techniques can be applied to the identified factors. Here again it may be questionable whether sufficient data are available for doing robust factor identifications.

with the standard deviations $\sigma_{k,t}$ and $\sigma_{l,t}$ taken from scalar GARCH models as described in chapter 4, section 2.3 and the correlation coefficients $\rho_{k,l}$ assumed to be constant. Consistent parameter estimates may be obtained by first estimating the scalar GARCH models and then determining the correlation coefficients through the relationship:

$$\rho_{k,l} = \frac{1}{t_2 - t_1 + 1} \sum_{t=t_1}^{t_2} \frac{\left(u_{k,t} - \bar{u}_k\right)\left(u_{l,t} - \bar{u}_l\right)}{\hat{\sigma}_{k,t}\hat{\sigma}_{l,t}} \tag{7-9}$$

It is thereby necessary to chose the time segments t_1 and t_2 in such a way that the relative price changes $u_{k,t}$ and $u_{l,t}$ are defined in the whole interval. To satisfy this condition for all product pairs (k, l), the starting point t_1 has to be chosen taking into account the time series with the lowest data availability, i.e. mostly the monthly peak product with the longest time to delivery. In principle, one could also choose the interval $[t_1\ t_2]$ differently for each product pair (k, l) to use as far as possible the available information on correlations. Yet the computed correlation matrix might not be positive semi-definite in this case, which gives rise both to theoretical inconsistencies and practical problems.

A further practical difficulty may arise if the price changes are not exactly distributed according to a GARCH model specification: the diagonal elements in the correlation matrix can then be different from 1, raising again theoretical and practical difficulties. A pragmatic, albeit theoretically not very well-founded remedy is to make an a posteriori adjustment using the formula:

$$\tilde{\rho}_{k,l} = \frac{\rho_{k,l}}{\sqrt{\rho_{k,k}\rho_{l,l}}} \tag{7-10}$$

This corresponds to pre- and postmultiplying the estimated correlation matrix by the diagonal matrix $< \rho_{k,k}^{-1/2} >$. Alternatively, one might envisage disregarding the GARCH model completely when determining the correlations, using the standard correlation formula:

$$\rho_{k,l} = \frac{1}{(t_2 - t_1 + 1)\bar{\sigma}_{k,t}\bar{\sigma}_{l,t}} \sum_{t=t_1}^{t_2} \left(u_{k,t} - \bar{u}_k\right)\left(u_{l,t} - \bar{u}_l\right) \tag{7-11}$$

This sets the diagonal elements of the correlation matrix automatically to one. Furthermore, higher correlations are usually determined than with the original approach. This is due to the fact that days with strong price changes are only entering the original correlation computation with a weight of $1/(\hat{\sigma}_{k,t}\hat{\sigma}_{l,t})$, in the second case the weight is however $1/(\overline{\sigma}_{k,t}\overline{\sigma}_{l,t})$. Since both the estimated volatility and the correlations are usually higher on days with strong price changes, the first variant will provide a lower weighting of these extreme events (and consequently lower volatility coefficients) than the second. Thus the second variant is more robust for risk measurement purposes, since here the focus is on extreme events.

In Table 7-1, results of selected parameter estimates using a GARCH (1,1) model on German forward price quotes are given. The GARCH and ARCH terms differ from zero for most products and the correlations among electricity products are also rather important. By contrast, the correlations between coal and electricity are found to be rather low.

Table 7-1. Parameter estimates for GARCH models for forward price time series

Product	Mean Variance σ_m^2	Impact previous changes α	Impact previous variance β	Log-Likelihood
Base year 03	0.0001159	0.334	0.000	1525.6
Peak year 03	0.0001312	0.043	0.956	1597.8
Base month 1202	0.0000447	0.532	0.216	694.7
Peak month 1202	0.0000496	0.110	0.861	681.4
Coal year 03	0.0000515	0.000	0.000	915.2

Correlations	Base year 03	Peak year 03	Base month 11-02	Peak month 11-02	Coal year 03
Base year 03	1.000	0.746	0.471	0.440	0.068
Peak year 03	0.746	1.000	0.475	0.448	0.101
Base month 11-02	0.471	0.475	1.000	0.564	0.147
Peak month 11-02	0.440	0.448	0.564	1.000	0.068
Coal year 03	0.068	0.101	0.147	0.068	1.000

Source: Own calculations, using forward price data until 1st October 2002

In order to assess the quality of the model, the implied kurtosis (the fourth moment of the distribution, which describes whether tails and peak of the distribution are more pronounced than in a normal distribution) is computed for the unconditional distribution resulting of the estimated GARCH-model. This kurtosis is compared to the kurtosis in the sample. As shown in Table 7-2, the observed kurtosis is much higher than the one implied by the GARCH model. This is a strong indication that the GARCH model with its conditional normal distribution is not fully suited to model the price changes on the forward market.

Besides the theoretical lack of consistency, this a second major reason for looking for alternatives involving mixtures of normal distributions and regime switching which will be discussed in the next section.

Table 7-2. Unconditional kurtosis of price changes implied by parameter estimates for GARCH models compared to actual observations

Product	Observed excess kurtosis	Implied excess kurtosis, GARCH model
Base year 03	24.09	1.00
Peak year 03	25.26	0.82
Base month 11-02	13.41	1.31
Peak month 11-02	8.44	0.03
Coal year 03	4.87	0.00

4.2   Multivariate models with regime switching

As with GARCH models, one might envisage also to specify explicitly multivariate extensions to switching regime models[52]. Yet the problems encountered with multivariate GARCH specifications arise similarly or even in an increased form in any attempt to model multivariate switching regimes. The number of parameters rapidly gets prohibitively large, the system-wide estimation may easily produce nonsense results for ill-specified models and the use and interpretation of correlation coefficients is subject to multiple precautions. On the last point, one might note that the concept of correlation coefficients fits well with multivariate normally distributed errors, since the distribution parameters include the variance-covariance matrix, of which the correlation matrix is the normalized form. But for non-normal distributions, no such obvious connection between correlation matrix and multivariate model specification exists.

A way to overcome this dilemma in a theoretically consistent way is to use the concept of copula (cf. Nelsen 1999, Embrechts et al. 2000, Lindskog 2000). Copula are defined as functions $\Gamma(u_1, \ldots, u_n)$ which establish the link between a general multivariate distribution $F_x(x_1,\ldots,x_n)$ of multiple stochastic variables $\mathbf{x} = [x_1,\ldots,x_n]^T$ and the univariate (or more properly: marginal) distributions of the single variables $F_1(x_1)$, $F_2(x_2)$, … $F_n(x_n)$. Note that the distributions are described here throughout through their cumulative distribution functions F_x, and not their probability density functions f_x. Formally, we can write:

$$F_k : \begin{matrix} \mathfrak{R} \\ x_k \end{matrix} \begin{matrix} \mapsto \\ \to \end{matrix} \begin{matrix} [0,1] \\ F_k(x_k) \end{matrix} \tag{7-12}$$

---

[52]   Examples of multivariate mixture models (not switching regimes) are discussed in Snoussi and Mohammad-Djafari (2000).

$$F_{\mathbf{x}} : \begin{array}{ccc} \Re^n & \mapsto & [0,1] \\ \mathbf{x} & \rightarrow & F_{\mathbf{x}}(\mathbf{x}) \end{array} \tag{7-13}$$

Then the copula Γ is defined as:

$$\Gamma : \begin{array}{c} [0,1]^n \mapsto [0,1] \\ \mathbf{u} \rightarrow \Gamma(\mathbf{u}) \therefore F_{\mathbf{x}}(x_1, x_2, ..., x_n) = \Gamma(F_1(x_1), F_2(x_2), ...F_n(x_n)) \end{array} \tag{7-14}$$

Hence the definition of Γ is an implicit one, as the function linking (coupling) the marginal distributions F_k to the overall distribution $F_{\mathbf{x}}$. Sklar's theorem (cf. Sklar 1996) proves that the copula Γ exists for any given $F_{\mathbf{x}}$ and moreover is unique, if all the marginal distributions F_k are continuous[53].

For practical purposes, it is most interesting to use the concept of copula the other way round: in addition of the specification for the marginal distributions F_k we can specify the type of the copula Γ and then have a nested approach for approximating the general multivariate distribution $F_{\mathbf{x}}$.

Several choices are possible for the copula, including the Farlie-Gumbel-Morgenstern family of copula, the Marshall-Olkin family or t-copulas (cf. Nelsen 1999, Lindskog 2000). We will however limit ourselves to the so-called Gaussian copula Γ_{Ga} which provides a natural extension of the concept of correlation to non-normal multivariate distributions. Further research is required to assess the usefulness of other copula definitions and to determine the relative strengths and weaknesses.

The Gaussian copula Γ_{Ga} is defined for any set of marginal distributions $F_k(x_k)$ through:

$$\Gamma_{Ga}(F_1(x_1), F_2(x_2), ...F_n(x_n)) = \phi_{\Sigma}\left(\phi_1^{-1}(F_1(x_1)), \phi_1^{-1}(F_2(x_2)), ..., \phi_1^{-1}(F_n(x_n))\right) \tag{7-15}$$

ϕ_1 is thereby the cumulative distribution of the standard normal distribution whereas ϕ_{Σ} is the cumulative distribution of the multivariate standard normal distribution with correlation matrix Σ and mean 0. The Gaussian copula is thus obtained by retransforming ("redistributing") the cumulative marginal distributions F_k into standard normal distributions using the inverse ϕ_1^{-1} and applying on these distributions the correlation matrix Σ. Estimates for the elements of Σ may be obtained by first estimating the

---

[53] Even if F_k is not continuous, the copula is at least unique on the range of values taken by F_k.

parameters of the marginal distributions $\hat{F}_1(x_1)...\hat{F}_n(x_n)$, then computing the transformed values $\tilde{x}_{k,t} = \phi^{-1}(\hat{F}_k(x_{k,t}))$ and finally estimating the correlation matrix $\hat{\Sigma}$ from the transformed $\tilde{x}_{k,t}$.

The most straightforward way of applying Gaussian copula to switching-regime models is to compute them directly for the daily relative price changes u_t. Yet this may lead to a substantial underestimation of the risk if the longer term dependence between price changes for different products is larger than the day-to-day dependence. This is obviously the case as illustrated through the examples of correlations of daily changes and 10-day-changes given in Table 7-3.

For the short-term risk with non-negligible holding period it is thus preferable to estimate the joint distribution for the price changes over the holding period. The distribution of the price changes for each stochastic variable is thereby estimated on the basis of the daily price changes, but the copula describing the dependence between the different stochastic variables are estimated based on the price changes during the holding period.

Thus the price changes $u_{k,t}$ for a product traded on the forward market are written as:

$$u_{k,t} \sim M_{k,t}\left(Pr_{k,0}, T_{k,11}, T_{k,22}, \mu_{k,1}, \sigma_{k,1}, \mu_{k,2}, \sigma_{k,2} \big| s_{k,t-1}\right) \quad (7\text{-}16)$$

Table 7-3. Correlations of relative price changes for forward price time series, depending on the time interval considered

Correlations 1-day price changes	Base year 03	Peak year 03	Base month 11-02	Peak month 11-02	Coal year 03
Base year 03	1.000	0.779	0.468	0.415	0.043
Peak year 03	0.779	1.000	0.468	0.472	0.115
Base month 11-02	0.468	0.468	1.000	0.597	0.145
Peak month 11-02	0.415	0.472	0.597	1.000	0.149
Coal year 03	0.043	0.115	0.145	0.149	1.000
Correlations 10-day price changes	Base year 03	Peak year 03	Base month 11-02	Peak month 11-02	Coal year 03
Base year 03	1.000	0.815	0.591	0.641	0.156
Peak year 03	0.815	1.000	0.516	0.679	0.092
Base month 11-02	0.591	0.516	1.000	0.748	0.431
Peak month 11-02	0.641	0.679	0.748	1.000	0.180
Coal year 03	0.156	0.092	0.431	0.180	1.000

$M_{k,t}$ stands there for a mixture of normal distributions with probabilities $Pr_{k,0}$, $Pr_{k,1,t}$ and $Pr_{k,2,t}$. As in chapter 4, section 2.3.4, the first state with probability $Pr_{k,0}$ corresponds to no price change at all, whereas the states 1 and 2 with probabilities $Pr_{k,1,t}$ and $Pr_{k,2,t}$ correspond to normal and extreme price changes with mean $\mu_{k,1}$ resp. $\mu_{k,2}$ and standard deviation $\sigma_{k,1}$ resp. $\sigma_{k,2}$. In order to derive a parsimonious model, the probabilities $Pr_{k,0}$ are assumed to be independent of the state and price change at the previous time step $t-1$. Thus the model used for describing the forward price changes is a mixture of no price change and price change, with a nested switching regime model for the part dealing with price changes. The a priori probability density function for $u_{k,t}$ conditional on the previous state $s_{k,t-1}$ then writes:

$$f_{u_{k,t}}\left(u_{k,t}|s_{k,t-1}\right) = Pr_{k,0}\delta_0\left(u_{k,t}\right)+$$
$$\left(1-Pr_{k,0}\right)\left(\Pr\left(s_{k,t}=1|s_{k,t-1}\right)f_N\left(u_{k,t};\mu_{k,1},\sigma_{k,1}\right)+\right. \quad (7\text{-}17)$$
$$\left. \Pr\left(s_{k,t}=2|s_{k,t-1}\right)f_N\left(u_{k,t};\mu_{k,2},\sigma_{k,2}\right)\right)$$

Using the description of Markov switching from equation (4-51), the a priori probabilities may be written:

$$\begin{bmatrix} \Pr\left(s_{k,t}=1|s_{k,t-1}\right) \\ \Pr\left(s_{k,t}=2|s_{k,t-1}\right) \end{bmatrix} = \begin{bmatrix} T_{k,11} & 1-T_{k,22} \\ 1-T_{k,11} & T_{k,22} \end{bmatrix}\begin{bmatrix} \Pr\left(s_{k,t-1}=1\right) \\ \Pr\left(s_{k,t-1}=2\right) \end{bmatrix} \quad (7\text{-}18)$$

The a posteriori probabilities $\Pr(s_{k,t-1}=j)$ can be obtained by comparing the probability density for the observation $u_{k,t-1}$ given state j to the overall probability density of the observation. For the state $s_{k,t}$ given the observation $u_{k,t}$ this writes:

$$\Pr\left(s_{k,t}=j|u_t\right)=$$
$$\frac{\Pr\left(s_{k,t}=j|s_{k,t-1}\right)f_N\left(u_{k,t};\mu_{k,j},\sigma_{k,j}\right)}{\Pr\left(s_{k,t}=1|s_{k,t-1}\right)f_N\left(u_{k,t};\mu_{k,1},\sigma_{k,1}\right)+\Pr\left(s_{k,t}=2|s_{k,t-1}\right)f_N\left(u_{k,t};\mu_{k,2},\sigma_{k,2}\right)}$$
$$(7\text{-}19)$$

At given parameters, equations (7-18) and (7-19) allow to compute recursively the probability of $s_{k,t}=j$ for each time step. This is the so-called Hamilton filter (cf. Hamilton 1989, Kim, Nelson 1999).

The log-likelihood function to be maximized for each product k is then (cf. equation (4-54)):

$$LL\left(Pr_{k,0}\,T_{k,11},T_{k,22},\mu_{k,1},\sigma_{k,1},\mu_{k,2},\sigma_{k,2}\right)=\sum_{t}\ln\left(f_{u_{t}}\left(u_{k,t}\big|s_{k,t-1}\right)\right)$$

$$=\sum_{t}\ln\left(Pr_{k,0}\delta_{0}\left(u_{k,t}\right)+\left(1-Pr_{k,0}\right)\left(f_{1}\left(u_{k,t};\mu_{k,1},\sigma_{k,1}\right)Pr\left(s_{k,t}=1\big|s_{k,t-1}\right)+\right.\right.$$

$$\left.\left.f_{2}\left(u_{k,t};\mu_{k,2},\sigma_{k,2}\right)Pr\left(s_{k,t}=2\big|s_{k,t-1}\right)\right)\right)$$

$$(7\text{-}20)$$

with the probabilities given through (7-18) and (7-19). Empirical results for the estimation of these parameters are given in Table 7-4. Obviously, most probabilities are significant and there is a significant difference in volatility between the two regimes with normal and high price changes. The probabilities for high price changes can be computed from the transition matrix probabilities and are found to range between 0.02 and 0.3.

Table 7-4. Parameter estimates for switching regime models for forward price time series

Product	$T_{k,11}$	$T_{k,22}$	$\mu_{k,1}$	$\mu_{k,2}$	$\sigma_{k,1}$	$\sigma_{k,2}$	$Pr_{k,0}$	LL_k
Base year 03	0.906	0.772	0.000	-0.002	0.004	0.020	0.033	1717.3
Peak year 03	0.957	0.807	0.000	-0.006	0.004	0.029	0.062	1775.1
Base month 11-02	0.988	0.528	0.001	0.011	0.007	0.031	0.204	723.2
Peak month 11-02	0.899	0.677	0.001	0.003	0.006	0.017	0.209	710.0
Coal year 03	0.915	0.000	0.000	-0.011	0.006	0.018	0.189	524.5

Source: Own calculations, using forward price data until 1st October 2002

Based on these estimates, the cumulative distribution function for price changes over the holding period d_h may be computed. Given that the distribution function for the one day price changes is always a mixture of normal distributions, the sum over price changes of several days will also follow a mixture of normal distributions. However, if each distribution is a mixture of three normals, a sum of d_h price changes has to be described by $d_h(d_h-1)/2$ terms, since within d_h days the first regime may occur l_1 times, with l_1 less or equal to d_h, the second regime may then occur up to d_h-l_1 times and the frequency of the third is given by $d_h-l_1-l_2$. The probabilities of each regime at each time step can be computed using equation (7-18). The probability $Pr_{tot,k}(l_1,l_2,t)$ for l_1 times regime 1 and l_2 times regime 2 after t time steps is then determined recursively through:

$$Pr_{tot,k}\left(l_1,l_2,t\right)=Pr_{tot,k}\left(l_1,l_2,t-1\right)\cdot Pr\left(s_{k,t}=0\big|s_{k,t-1}\right)$$

$$+1_{l_1>0}\,Pr_{tot,k}\left(l_1-1,l_2,t-1\right)\cdot Pr\left(s_{k,t}=1\big|s_{k,t-1}\right)$$

$$+1_{l_2>0}\,Pr_{tot,k}\left(l_1,l_2-1,t-1\right)\cdot Pr\left(s_{k,t}=2\big|s_{k,t-1}\right)$$

$$(7\text{-}21)$$

$$Pr_{tot,k}\left(0,0,0\right)=1$$

For the mean and standard deviation of this combination of regimes we get according to the general rules for addition of independent normal distributions:

$$\mu_{tot,k}(l_1,l_2,t)=l_1\mu_{k,1}+l_2\mu_{k,2}$$
$$\sigma_{tot,k}(l_1,l_2,t)=\sqrt{l_1\sigma_{k,1}^{2}+l_2\sigma_{k,2}^{2}}$$

(7-22)

Hence the distribution function for $u_{tot,k,t}=\sum_{\tau=t+1}^{t+d_h}u_{k,\tau}$ may be written:

$$F_{tot}(u_{tot,k,t})=\sum_{l_1=0}^{d_h}\sum_{l_2=0}^{d_h-l_1}\Pr_{tot}(l_1,l_2,d_h)F_N(u_{tot,k,t};\mu_{tot}(l_1,l_2,d_h),\sigma_{tot}(l_1,l_2,d_h))$$

(7-23)

Consequently, all observations $u_{tot,k,t}$ for the different products k may be transformed to drawings $\widetilde{u}_{tot,k,t}$ from a standard normal distribution using

$$\widetilde{u}_{tot,k,t}=\phi^{-1}(F_{tot}(u_{tot,k,t}))$$

(7-24)

From these drawings, the correlation matrix Σ characterizing the Gaussian copula Γ_{Ga} is then computed using the simplified formula

$$\Sigma_{kl}=\frac{1}{t_2-t_1+1}\sum_{t=t_1}^{t_2}\widetilde{u}_{tot,k,t}\widetilde{u}_{tot,l,t}$$

(7-25)

given that the standard deviation of $\widetilde{u}_{tot,k,t}$ and $\widetilde{u}_{tot,l,t}$ is equal to 1 by construction. For the products mentioned in Table 7-4, the copula correlation coefficients are given in Table 7-5.

Table 7-5. Estimates of coefficients for Gaussian copula for forward price changes on a 10-day interval

Copula	Base year 03	Peak year 03	Base month 11-02	Peak month 11-02	Coal year 03
Base year 03	1.000	0.815	0.602	0.647	0.151
Peak year 03	0.815	1.000	0.511	0.681	0.095
Base month 11-02	0.602	0.511	1.000	0.721	0.392
Peak month 11-02	0.647	0.681	0.721	1.000	0.152
Coal year 03	0.151	0.095	0.392	0.152	1.000

Source: Own calculations, using forward price data until 1st October 2002

Obviously there is a strong positive dependence between the different electricity products whereas the interdependence with coal products is rather limited. The calculated copula coefficients are also found to be rather similar to the 10-day-correlations given in Table 7-3.

5.   POWER PLANT MODELS FOR RISK QUANTIFICATION

Besides price models, models for characterizing the optionalities and the quantity risks inherent in power plants , are also needed for assessing the integral earnings at risk. Of course, the real option models presented in chapter 6 may be used for this purpose, at least as far as the delivery risk is concerned. But some adaptation is required as discussed in section 5.1 together with a simplified representation of CHP plants for risk valuation purposes. Section 5.2 then derives a workable extension for the evaluation of the short-term risk, which does not assume that forward and spot prices follow the same distribution. This approach then allows also analyzing whether the "expected operation hedge" introduced in chapter 6, section 5 is a delta hedge or not.

5.1   Delivery risk – including analytical model for CHP valuation

The real option models described in chapter 6, section 3 allows computing the value of any thermal power plant without cogeneration treated as real option. For risk management purposes it is however not sufficient to compute only the expected value, but rather the distribution of the power plant value derived from the distribution of the spot prices is needed. Given that anyhow the computation of the option value requires computing the value recursively each day for different price scenarios (cf. chapter 6, section 3), the computation of the value distribution is mostly a question of tracking the distributions over time, taking into account the information on transition probabilities between price scenarios of different days. With 10 or more price scenarios per day, the final distribution may however not be obtained by enumerating and ordering all possible combinations of daily price patterns, since already for one month and 10 price scenarios per day a total of 10^{30} paths and corresponding power plant values is obtained. Instead a numeric convolution (folding) of the distributions has to be applied, which allows determining recursively the probability for the power plant value to lie within pre-specified intervals.

Besides the power plant value, also the electricity production and fuel consumption can be tracked in this way.

For CHP systems, the cogeneration of heat and power induces further complexity not only for operation but also for risk management. Firstly, the heat demand obviously induces a quantity risk, which can usually not be hedged on a heat forward or future market. However, one may envisage hedging at least for part of the heat demand risk through weather derivatives (cf. e.g. Simpson 1997, Ali 2000, Kraus 2002). This will not provide a perfect hedge, since part of the heat demand fluctuations are related to other factors such as changes in production. But if the CHP plant is delivering to a municipal district or local heating grid, much of the demand will be dependent on the outside temperature and can be hedged, if a counterpart is willing to take over this risk.

But cogeneration does not only introduce an additional risk factor, it also affects the power plant operation and the electricity output. For systems like back-pressure turbines, gas turbines without auxiliary cooler or motor engine CHP, the electricity output is directly coupled to the heat output (cf. chapter 6, section 1). The operation of these plants as stand-alone systems is fully determined by the heat demand pattern and therefore no optionality is left[54]. Extraction-condensing turbines mentioned in chapter 6, section 1.4 and larger CHP systems are more complex options that require specific models. In the following, an analytical approximation for CHP systems with extraction-condensing turbines is proposed in order to circumvent the use of the full stochastic optimization model described in chapter 6, section 4 when it comes to risk management.

The operation of an extraction-condensing CHP plant is limited through several restrictions as depicted graphically in Figure 6-1 and Figure 7-7. If the heat output $\dot{Q}_{u,t,s}^{HT}$ is given, then also the electricity output $P_{u,t,s}^{EL}$ has to surpass a certain minimum level, given either by the minimum steam production or the minimum electricity-to-heat ratio. Both the heat and the minimum electricity quantity are fixed and from a risk management point of view, they should be rather treated as forwards $\dot{Q}_{FW}^{HT}$ and P_{FW}^{EL} than as options. Yet another part of the electricity still is optional. It is the quantity P_{Opt}^{EL} between the lower load limit and the upper load limit in the P-Q chart (cf. Figure 7-7). This part should be treated as an option, with the amount of the option obviously depending on the heat load. Correspondingly there is

---

[54]   Of course, as soon as these plants are combined with a storage or a peak load device, some optionality for plant operation is given and may be considered.

also an optional part P_{Opt}^{FU} and a fixed forward P_{FW}^{FU} for the fuel consumption.

If several extraction-condensing turbines with different plant characteristics are available to satisfy a given heat demand, the optimal dispatch has to be modeled at least approximately in any analytical risk management tool. This dispatch will in principle be done in the order of increasing heat production costs. Yet opportunity costs for electricity have thereby to be taken into account. These are positive if the marginal costs for the electricity produced are above the market prices. In this case, the difference between marginal electricity generation costs and market prices has to be attributed to the heat generation, penalizing those units with high minimum power to heat ratio, where much electricity is produced in "back-pressure mode". On the other hand, if marginal electricity generation costs are below market prices, the units can earn money by producing solely electricity. Thus opportunity costs arise for any reduction in the electricity output due to the heat generation. These have to be added to the heat generation costs proportionally to the marginal electricity reduction rate. The units with high marginal electricity reduction will hence have higher marginal heat costs and will consequently be rather used solely for electricity generation.

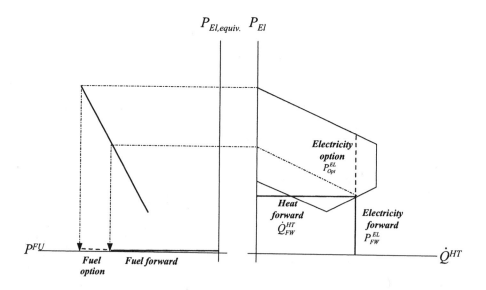

Figure 7-7. Analytical model for extraction-condensing CHP plants

In summary, the marginal heat costs to be taken as basis for the heat dispatch decision may be written:

$$c_{u,t,s}^{HT} = \beta_u^{HT} p_{f,t,s}^{FU} + 1_{p_{t,s}^{EL} < \beta_u^{EL}} s_{u,min}^{EL} \left(\beta_u^{EL} p_{f,t,s}^{FU} - p_{t,s}^{EL} \right)$$
$$+ 1_{p_{t,s}^{EL} > \beta_u^{EL}} \frac{\beta_u^{HT\,\prime}}{\beta_u^{EL\,\prime}} \left(p_{t,s}^{EL} - \beta_u^{EL} p_{f,t,s}^{FU} \right) \tag{7-1}$$

Within the Monte-Carlo simulation of delivery risk, the dispatch decision has then to be simulated. This is somewhat more complicated than describing the exercise of a simple spark spread option, but it is still simpler than applying the full stochastic optimization model.

5.2 Short-term risk

Besides the delivery risk also the short-term risk for a given portfolio has to be determined for the integral earnings at risk. The short-term risk corresponds to changes in the value (or rather the expected earnings) of the assets in the portfolio due to changes in the market prices. Obviously for market traded products like forwards, the value change corresponds to changes in the forward prices. But for power plants, the calculation of the short-term price risk is more difficult. In fact, it describes the impact of changes of forward prices on the power plant exercise during the delivery period and the corresponding earnings. Its computation requires therefore first the selection of a model describing the link between spot and forward prices (cf. section 4). If this has been done, the value of the real option has in principle to be computed for each possible change in the forward price. This is clearly not feasible. Yet one can safely assume (even if here no formal proof is given) that the option value will be a smooth, i.e. continuous and (twice) differentiable, function G of the forward price. Then changes in the option value for arbitrarily forward price changes can be computed from a second order Taylor expansion of the value function G:

$$G_{T,s} = G\left(p_{T,s}^{FU}, r_{T,s}^{OP/FU}, r_{T,s}^{PE/FU} \right)$$
$$= V_0 + \left. \frac{\partial G}{\partial p^{FU}} \right|_{p_0} \left(p_{T,s}^{FU} - p_0^{FU} \right) + \left. \frac{\partial G}{\partial r^{OP/FU}} \right|_{p_0} \left(r_{T,s}^{OP/FU} - r_0^{OP/FU} \right)$$
$$+ \left. \frac{\partial G}{\partial r^{PE/FU}} \right|_{p_0} \left(r_{T,s}^{PE/FU} - r_0^{PE/FU} \right) + \frac{1}{2} \left. \frac{\partial^2 G}{\partial p^{FU2}} \right|_{p_0} \left(p_{T,s}^{FU} - p_0^{FU} \right)^2 \tag{7-2}$$
$$+ \frac{1}{2} \left. \frac{\partial^2 G}{\partial r^{OP/FU2}} \right|_{p_0} \left(r_{T,s}^{OP/FU} - r_0^{OP/FU} \right)^2 + \frac{1}{2} \left. \frac{\partial^2 G}{\partial r^{PE/FU2}} \right|_{p_0} \left(r_{T,s}^{PE/FU} - r_0^{PE/FU} \right)^2$$

The value is expressed as a function of the fuel price $p_{T,s}^{FU}$ and the "spark-spread" ratios $r_{T,s}^{OP/FU}$ and $r_{T,s}^{PE/FU}$ of off-peak and peak electricity price to fuel price. This parameterization is chosen since fuel price changes, if not accompanied by changes in the spark-spread ratios will hardly affect the power plant operation (cf. chapter 6). By contrast changes in the spark-spread ratios directly affect the profitability of the plant operation and thus the option exercise. The partial derivative $\partial G/\partial p^{FU}$ with respect to the fuel price has therefore to be interpreted as the value alteration with respect to a general change in the (fuel and electricity) price level, whereas the partial derivatives with respect to the spark-spread ratios $\partial G/\partial r^{PE/FU}$ and $\partial G/\partial r^{OP/FU}$ correspond to the effect of electricity price changes at constant fuel prices. Besides these first order effects, which correspond to the "Deltas" in financial mathematics, the second order effects, the so-called Gammas, are also included. Thus the two most important risk factors from financial analysis are taken into account (cf. e.g. Hull 2000). An extension to other risk factors like volatility change risk ("Vega") is straightforward but rather cumbersome, since here again the multiple underlyings of the power plant option have to be tackled. In order to simplify the expression, the mixed second order derivatives have already been omitted in equation (7-2), since there is no obvious reason why the impact of the spark-spread ratios on the value should be modified by different fuel price levels or by the level of the other spark-spread ratio. These are defined for the time segments peak and off-peak instead of the products base and peak, since the power plants are also operated time segment wise.

The derivatives used in (7-2) can be computed numerically using evaluations of the value function at different price levels. For example, the derivative with respect to fuel prices is determined using besides the value for the reference case the option value obtained when varying all prices by a same increment Δp^{FU} upwards and downwards (e.g. +/- 20 % for each input parameter):

$$\frac{\partial G}{\partial p^{FU}}\bigg|_{P_0} = \frac{G\left(p_0^{FU}\left(1+\Delta p^{FU}\right),r_{t,s}^{OP/FU},r_{t,s}^{PE/FU}\right)-G\left(p_0^{FU}\left(1-\Delta p^{FU}\right),r_{t,s}^{OP/FU},r_{t,s}^{PE/FU}\right)}{2\cdot p_0^{FU}\Delta p^{FU}}$$

$$\frac{\partial^2 G}{\partial p^{FU2}}\bigg|_{P_0} = \frac{G\left(p_0^{FU}\left(1+\Delta p^{FU}\right),r_{t,s}^{OP/FU},r_{t,s}^{PE/FU}\right)-2\cdot G\left(p_0^{FU},r_{t,s}^{OP/FU},r_{t,s}^{PE/FU}\right)}{2\cdot p_0^{FU2}\Delta p^{FU2}}$$

$$+\frac{G\left(p_0^{FU}\left(1-\Delta p^{FU}\right),r_{t,s}^{OP/FU},r_{t,s}^{PE/FU}\right)}{2\cdot p_0^{FU2}\Delta p^{FU2}} \qquad (7\text{-}3)$$

Here central differences have been used to increase the computational accuracy. For the Taylor expansion, it is hence necessary to compute seven different scenarios of delivery values, namely the reference scenario and up- and down-variations for each of the three parameters. These computations may be done once a week or once a month and in between the approximations through the function (7-2) may be used when determining the short-term risk.

For this calculation within the framework of a Monte-Carlo simulation, the arguments of the value function have to be computed from the simulated set of forward prices $p_{T,s}^{FU}$, $p_{T,s}^{PE}$ and $p_{T,s}^{OF}$ using the expressions:

$$r_{T,s}^{PE/FU} = \frac{p_{T,s}^{PE}}{p_{T,s}^{FU}}$$

$$p_{T,s}^{OF} = \frac{p_{T,s}^{BA} \cdot d^{BA} - p_{T,s}^{PE} \cdot d^{PE}}{d^{BA} - d^{PE}} \qquad (7\text{-}4)$$

$$r_{T,s}^{OF/FU} = \frac{p_{T,s}^{OF}}{p_{T,s}^{FU}}$$

Then the value function (7-2) for the power plant(s) may be applied to determine the contribution of the power plant(s) within the overall combined trading and generation portfolio. By comparing the different Monte-Carlo simulations, the short-term risk can then be computed as usual as the conditional value at risk on the empirical distribution.

For the case of a simple spark-spread option, some analytical insights may be gathered on beforehand, highlighting the relation between optimal hedging and expected operation of the power plants. The value function is in this setting additively separable by time segments and, taking into account equations (7-7) and (7-4), it may be written:

$$G_{T,s} = \sum_{t_D \in T^{PE}} G_{t_D,s} + \sum_{t_D \in T^{OF}} G_{t_D,s}$$

$$= P_{max}^{EL} \sum_{t_D \in T^{PE}} \max\left(0, p_{t_D,s}^{PE} - \frac{1}{\eta} p_{T,s}^{FU}\right) + P_{max}^{EL} \sum_{t_D \in T^{OF}} \max\left(0, p_{t_D,s}^{OF} - \frac{1}{\eta} p_{T,s}^{FU}\right)$$

$$= P_{max}^{EL} p_{T,s}^{FU} \sum_{t_D \in T^{PE}} \max\left(0, r_{T,s}^{PE} k_{t_D,s} - \frac{1}{\eta}\right) + P_{max}^{EL} p_{T,s}^{FU} \sum_{t_D \in T^{OF}} \max\left(0, r_{T,s}^{OF} k_{t_D,s} - \frac{1}{\eta}\right)$$

$$(7\text{-}5)$$

The max operator describes thereby the condition that the plant is only operating if instantaneous electricity prices exceed fuel costs respectively if the current electricity-fuel-price ratio exceeds the heat rate.

Obviously the derivative of the value function with respect to the fuel price (or rather the general price level) will be independent of the fuel price level. More interesting are the derivatives with respect to the spark-spread ratios, here shown for the peak spark-spread ratio:

$$
\frac{\partial G_{T,s}}{\partial r_{T,s}^{PE}} = P_{max}^{EL}\, p_{T,s}^{FU} \sum_{t_D \in \mathbf{T}^{PE}} \left(1_{r_{T,s}^{PE} k_{t_D,s} - \frac{1}{\eta}} k_{t_D,s} \right)
$$

$$
\geq P_{max}^{EL}\, p_{T,s}^{FU} \sum_{t_D \in \mathbf{T}^{PE}} \left(1_{r_{T,s}^{PE} k_{t_D,s} - \frac{1}{\eta}} \right) \tag{7-6}
$$

The inequality in the second line states that the delta is larger than the expected operation time $\sum_{t_D \in \mathbf{T}^{PE}} \left(1_{r_{T,s}^{PE} k_{t_D,s} - \frac{1}{\eta}} \right)$ times the maximum output power P_{max}^{EL} (The multiplier $p_{T,s}^{FU}$ is only appearing since the price ratio instead of actual prices is considered). This inequality holds since the average value of the hourly price factor $k_{t_D,s}$ exceeds 1 for those hours when the plant is operating. According to equation (7-7) $k_{t_D,s}$ is 1 on average over all hours, but the power plant is obviously mostly operating in those hours with higher $k_{t_D,s}$ so that the weighted average is above unity.

This has an immediate implication for hedging: The "expected operation hedge" discussed in chapter 6, section 5 will not provide a proper delta hedge. This is true since the expected operation hedge corresponds exactly to the second line in equation (7-6). For obtaining a delta hedge, it is under this price model hence not sufficient to compute the expected operation of the power plant at current forward prices and to sell respectively purchase the corresponding quantities on the forward market. But this result is dependent on the price model used. It can be shown that under the additive price model proposed in equation (7-6).

6.   APPLICATION

The application of the models described before requires the set up of a numerical simulation toolbox. How this can be done is described in chapter 6.1, whereas chapter 6.2 describes briefly the portfolios analyzed. Finally, chapter 6.3 is devoted to simulation results for selected portfolios.

6.1   Computation methodology

In textbooks, three basic alternatives are often proposed for value at risk computation, being the analytical approach, historical simulation and Monte-Carlo simulation (e.g. Hull 2000). Given the complexity of the price processes modeled and the multitude of products to be considered, only Monte-Carlo simulation is feasible in the present setting. This is also reasonable given that for delivery risk calculations, notably under the profit-at-risk approach, Monte-Carlo simulations are generally used.

The overall algorithm to be applied is depicted in Figure 7-8. Since two risk components have to be determined for integral earnings at risk computations, namely short-term risk and delivery risk, also two series of Monte-Carlo simulations are needed.

- Monte-Carlo simulations for spot market prices in delivery period

- Selection of hedge for delivery period

- Computation of portfolio value and delivery risk for delivery period

- Scenario variations for spot prices corresponding to forward price changes

- Calculation of derivatives (delta, gamma) for option value approximation in forward simulation

- Monte-Carlo simulations for forward market prices and portfolio values in holding period

- Determination of front end risk and computation of integral earnings at risk

Figure 7-8.   General algorithm for the computation of integral earnings at risk

Since the valuation of the power plant options on the forward market depends on their value on the spot market, the price simulations for the spot market have to be carried out first. Then the hedge to be used for the delivery period has to be defined and the corresponding value, delta and delivery risk have to be determined.

Through scenario variations, the parameters for the approximate description of power plant optionalities in the forward model are determined. Then the Monte-Carlo simulation for the forward prices is carried out which provides directly information on the short-term risk. Combining this information with the delivery risk derived in step 3, the total earnings at risk are obtained.

6.2   Portfolio description

The power plant portfolios analyzed are the same as the one considered in chapter 6. For each of these portfolios, five different hedges are considered at the end of the forward market ("final hedge"):
- zero hedge,
- complete hedge,
- hedge corresponding to plant operation at current peak and off-peak prices,
- hedge corresponding to plant operation at expected hourly prices,
- hedge corresponding to average plant use computed under a full stochastic model,

The delivery risk is first evaluated for the example of immediate delivery in October 2002. Then the future delivery risk for similar portfolios in October 2003 is estimated, not only at current forward prices but also with fuel and/or electricity price changes on the forward market. The impact of these forward price changes on the portfolio value at delivery allows also assessing the option characteristics to be used when evaluating the short-term risk. This is then done using the regime switching model for forward prices described in section 4.2. Thereby besides the five aforementioned hedges also a delta hedge is numerically derived and the corresponding short-term risk quantified. Finally the integral earnings at risk are computed.

6.3   Results

In the following, first the results for the delivery risk are discussed, and then results for the short-term risk are given.

6.3.1   Delivery risk

In Table 7-6, the earnings and the corresponding delivery risk for the month of October 2002 are summarized for the two power plants described in chapter 6, section .6.1.1 as viewed on the first day of this month. Whereas the mean value of the earnings is about 5.8 Mill. € for the coal plant and 2.7 Mill. € for the gas plant, the 1 % quantile of the earnings distribution is at 4.1 Mill. € for the coal plant and 1.5 Mill. € for the gas plant. Note that the expected value of the earnings is considerably higher than the option values determined in chapter 6, section 6.2, which were calculated using directly the prices from the market simulation model. As indicated in Figure 4-10, the last forward quote for October 2002 was about 29 €/MWh for the base product, whereas the simulated price is about 21 €/MWh and the actual spot market price was about 23 €/MWh, so that the valuation using the equivalent martingale measure (i.e. consistent with the forward quote) yields considerably higher values.

Table 7-6. Earnings, delivery risk and quantity variation of unhedged power plant portfolios]

Period analyzed: October 2002	Coal-fired plant	Gas-fired plant
Earnings		
Mean value	5792 T€	2725 T€
1 % quantile	4052 T€	1498 T€
99 % quantile	7809 T€	4778 T€
Delivery Risk		
"Value at risk"	1740 T€	1227 T€
Conditional earnings-at-risk	2034 T€	1454 T€
Electricity production		
Mean value	384.7 GWh	209.3 GWh
1 % quantile	341.2 GWh	158.6 GWh
99 % quantile	398.5 GWh	252.1 GWh
Fuel consumption		
Mean value	839.7 GWh	368.1 GWh
1 % quantile	756.8 GWh	277.2 GWh
99 % quantile	901.5 GWh	445.4 GWh

Source: own calculations, based on prices available 1 October 2002

The conditional earnings-at risk are in this case 2.0 Mill. € for the coal plant and 1.5 Mill € for the gas plant, i.e. more than half of the earnings in the case of the gas-fired plant and more than one third in the case of the coal fired plant. The conditional earnings-at-risk exceed the 1 % quantile (labeled here with the traditional but misleading term "value at risk") by about 10 % to 15 %. This relatively small discrepancy is due to the fact that the earnings are limited by zero from below in the case of a power plant. For a non-hedged load portfolio the discrepancy is considerably higher given the possibility of extreme price spikes.

The variation in electricity production between the 1 % and the 99 % quantile is about +/-10 % in the case of the coal-fired plant and about +/-25 % for the gas-fired plant and also the variation in fuel consumption is much lower than the variations in earnings. This shows that the earnings, which are formed by the margin above generation costs, are much more sensitive to electricity price risk than the physical production.

Table 7-7 indicates how the earnings at risk in the delivery period can be reduced through hedging. All proposed hedges lead to a substantial decrease of the risk incurred. In the case of the coal power plant, the conditional earnings at risk can be reduced to less than 10 % of the expected earnings. For the gas-fired plant, the remaining uncertainty is somewhat higher but not exceeding the 20 % percent for the best hedge. The higher residual risk occurs because the gas plant is more often out of the money and the optionality of switching off is not perfectly captured when using forwards as hedges. The question which is the best hedging strategy in this incomplete market setting can not be answered unequivocally. For the coal power plant, the full hedge and the peak-/off-peak hedge are identical, since at average peak and off-peak prices the coal plant is in the money throughout. And this hedge performs better than the deterministic hourly or the stochastic hedge. For the gas plant, the deterministic hedge however performs best. This illustrates that the theoretical difficulties in deriving an optimal hedge in incomplete markets (cf. section 3) translate also in empirical difficulties for determining the optimal hedge without recurring to stochastic optimization.

Table 7-7. Earnings at risk - delivery risk for power plant portfolios with various hedges

Delivery risk for October 2002	Coal-fired plant	Gas-fired plant
Zero Hedge	2034 T€	1454 T€
Full Hedge	299 T€	665 T€
Peak/Off-Peak Hedge	299 T€	614 T€
Deterministic Hedge	324 T€	470 T€
Stochastic Hedge	314 T€	518 T€

Source: own calculations, based on prices available 1 October 2002

The stochastic hedge, which hedges the expected exercise probability of the option, is outperformed by the other two hedges. This seemingly surprising result can be explained by the fact that the stochastic hedge leads to the lowest variance for the sum of electricity sold or purchased on the spot market, but it is not the one which limits best the downside risk in earnings. The deviation may be attributed both to the skewed price distributions and to the gamma risk (risk of changing exercise probabilities) inherent to any option and which is particularly relevant here, given that no continuous hedging is possible.

The delivery risk varies with variations in the fuel prices and variations in other factors which affect future spot prices. This is illustrated in Table 7-8 for two types of variations:

- variations of fuel prices with subsequent changes in electricity prices,
- variations of electricity prices at constant fuel prices.

Table 7-8. Earnings at risk - delivery risks for power plant portfolios with deterministic hedges and expected electricity production and fuel consumption under varying fuel and electricity prices

Coal-fired plant, delivery period: October 2003	Expected Values			Delivery risk
	Earnings	Electricity output	Fuel input	
Current prices for October 2003	6407 T€	288 GWh	646 GWh	590 T€
Fuel price changes[a]				
General proportional decrease	5033 T€	321 GWh	723 GWh	418 T€
General proportional increase	7872 T€	274 GWh	626 GWh	627 T€
Increase gas/coal ratio	5489 T€	314 GWh	706 GWh	473 T€
Decrease gas/coal ratio	7407 T€	277 GWh	621 GWh	626 T€
Electricity price changes				
General proportional decrease	5972 T€	278 GWh	624 GWh	550 T€
General proportional increase	6825 T€	298 GWh	668 GWh	591 T€
Increase peak/off-peak ratio	6675 T€	281 GWh	631 GWh	567 T€
Decrease peak/off-peak ratio	6125 T€	290 GWh	667 GWh	580 T€
Gas-fired plant, delivery period: October 2003	Expected Values			Delivery risk
	Earnings	Electricity output	Fuel input	
Current prices for October 2003	4337 T€	160 GWh	288 GWh	831 T€
Fuel price changes[a]				
General proportional decrease	3935 T€	170 GWh	304 GWh	557 T€
General proportional increase	4416 T€	144 GWh	259 GWh	1093 T€
Increase gas/coal ratio	1521 T€	109 GWh	197 GWh	841 T€
Decrease gas/coal ratio	5651 T€	235 GWh	421 GWh	726 T€
Electricity price changes				
General proportional decrease	4017 T€	149 GWh	270 GWh	772 T€
General proportional increase	4652 T€	169 GWh	305 GWh	902 T€
Increase peak/off-peak ratio	4711 T€	160 GWh	291 GWh	851 T€

[a] with implied changes in electricity prices according to integral market price model
Source: own calculations, based on prices available 1 October 2002 (forward price October 2003 replaced by forward price 4th quarter 2003)

For the variations in fuel prices, the 1 % and 99 % quantiles of the fuel price simulations from chapter 4, section 3 are taken. Firstly simulations for the expected fuel and electricity price level in October 2003 are carried out. Then for all fuels the lower resp. the upper quantiles are taken and simulations of electricity prices are performed. Subsequently the mean value of the fuel prices are left unchanged but the gas price is increased and the coal price is decreased by the same factor in a way that the resulting gas-to-coal price ratio attains the 99 % quantile observed in the simulations. Finally the opposite is done, decreasing the gas and increasing the coal price. In this case, the ratio decreases to 1.5 whereas in the opposite case it increases to 3.1. These two scenarios affect the optimal dispatch of the power plants and consequently their option value in principle more seriously than general increases or decreases. For the gas-fired plant consequently variations in the expected electricity output by more than a factor two are observed. But as indicated in Table 7-8, the impact is more limited in the case of the coal power plant.

The hedge used is the optimal deterministic hedge at the beginning of the delivery period. Consequently the delivery risk corresponds to the risk of deviations of observed electricity spot prices from expected forward prices at given fuel prices. Nevertheless the delivery risk varies by about a factor 1.5 for the coal-fired plant and almost a factor two for the gas-fired plant between the different fuel price scenarios. Calculating the delivery risk at given forward price levels will hence only provide a first idea of what the potential risk is and it is clearly preferable to determine the upper bound of the delivery risk by doing Monte-Carlo simulations on fuel and electricity forward prices.

The expected input and output quantities are subject to variations by about +/-10 % in the case of the coal fired plant and about +/-30 % for the gas-fired plant. When comparing the results of Table 7-8 with the results for October 2002 contained in Table 7-6, one may note that the coal-fired plant is operating much less in 2003 given that electricity prices are lower and fuel prices are higher than in 2002. In fact the coal plant is almost operating as a base load plant at the price level given for October 2002.

The variations in electricity prices used for the results in Table 7-8 are taken from forward price simulations over the holding period, chosen as 10 days. Again the 1 % and 99 % quantiles are applied and two types of price changes are distinguished: level changes, i.e. changes in the electricity base price and shape changes, i.e. changes in the peak-to-off-peak price ratio at constant average prices. As indicated in Table 7-8, the impact of these price risks on the expected input and output quantities is smaller. But this is mainly due to the fact that here only changes over the holding period are considered, whereas the fuel price risk is taken over a whole year.

For the CHP system described in chapter 6, section 6.1.2, the delivery risk for the unhedged portfolio can be obtained as a result of the stochastic operation optimization. To do so, a convolution of the distributions of the daily objective values, taking into account the transition probabilities has to be computed. The corresponding results are displayed in Table 7-9.

Here the unhedged earnings at risk are about 25 % of the expected value. Also variations in electricity production and fuel consumption are less pronounced than for the non-CHP power plants, given that part of the electricity produced has no optionality but is rather a direct consequence of the heat demand. The same holds for the fuel consumption.

Table 7-9. Earnings, delivery risk and quantity variation of unhedged CHP system portfolio

Period analyzed: October 2002	CHP system		
Earnings			
Mean value	4823 T€		
1 % quantile	3600 T€		
99 % quantile	6281 T€		
Delivery Risk			
"Value at risk"	1225 T€		
Conditional earnings-at-risk	1334 T€		
	Electricity production	*Coal consumption*	*Gas consumption*
Mean value	498 GWh	1272 GWh	160 GWh
1 % quantile	467 GWh	1189 GWh	152 GWh
99 % quantile	524 GWh	1339 GWh	178 GWh

Source: own calculations, based on prices available 1 October 2002

6.3.2   Short-term risk

Variations of the value at delivery as those obtained in the previous section for changes in fuel and electricity prices are taken as basis for computing the approximate option values entering the model for short-term risk. However, given that the short-term risk reflects changes in portfolio value due to short-term forward price changes, the deltas and gammas are calculated not based on expected fuel price changes over one year and corresponding electricity price changes, but rather based on fuel price changes which might occur within the holding period. Unfortunately, no reliable data on gas price volatility for Germany were available, so that coal price volatility has been applied instead. The results given in Table 7-10 indicate that the short-term risk is of the same order of magnitude than the delivery risk of a hedged power plant portfolio. Again the unhedged risk of the coal-fired plant is higher in absolute terms than the one of the gas-fired plant, but relatively to the expected earnings, the risk is higher for the gas fired plant. The full hedge is obviously inefficient and also a hedge based on a simple comparison of forward quotes for peak and off-peak with the strike price of the option does not improve the risk situation much. A hedge based on the expected power plant exercise leads to a reduction of the risk by about a factor of two, with only slight differences between the deterministic and the stochastic operation hedge. The delta hedge does considerably better at least for the gas-fired plant, leading to a short-term risk which is about a factor four lower than for an unhedged portfolio. The quantities corresponding to these hedges are summarized in Table 7-11. Whereas the Peak/Off-Peak hedge corresponds to the full hedge for the coal-fired plant, it is a pure peak hedge in the case of the gas-fired plant, given its higher operation cost. The deterministic expected operation and the stochastic expected operation differ roughly by about 10 %. The differences are considerably higher compared to the delta hedge. Here the quantities purchased for hedging are considerably higher in the case of the gas fired plant compared to the expected operation hedges. For the coal-fired plant the picture is less clear-cut. Peak electricity is purchased much more than under the operation based hedges, but for base electricity and fuel the opposite is observed. This can be attributed to indirect effects of the fuel prices on the parameters of the value function and on the consequences of correlations between price changes.

Table 7-10. Earnings at risk – short-term risk for power plant portfolios with various hedges

Short-term risk for October 2003 [T€]	Coal-fired plant	Gas-fired plant
Zero hedge	395	265
Full hedge	401	621
Peak/Off-Peak hedge	401	132
Deterministic expected operation hedge	207	127
Stochastic expected operation hedge	232	128
Delta hedge	181	72

Source: own calculations, based on prices available 1 October 2002

Table 7-11. Earnings at risk – hedging quantities for power plants

Hedging quantities for October 2003 [GWh]	Electricity -Base	Electricity -Peak	Fuel
Coal-fired plant			
Full hedge	446	0	997
Peak/Off-Peak hedge	446	0	997
Deterministic expected operation hedge	223	63	640
Stochastic expected operation hedge	201	57	573
Delta hedge	164	112	524
Gas-fired plant			
Full hedge	446	0	804
Peak/Off-Peak hedge	0	166	298
Deterministic expected operation hedge	54	59	204
Stochastic expected operation hedge	59	54	204
Delta hedge	80	96	244

Source: own calculations, based on prices available 1 October 2002

For the CHP system described in chapter 6, section 6.1.2, the short-term risk is given in Table 7-12. It is by about a factor of 1.5 higher than the risk of the coal power plant. But through a delta hedge it may be reduced to about one quarter of the maximum value. Again the delta hedge performs considerably better than the deterministic operation hedge. The corresponding hedging quantities are also given in Table 7-12. The delta hedge has a considerably higher peak quantity as compared to the other hedges, whereas the quantities contracted in the base period are substantially lower. A stochastic expected operation hedge is not computed here, since this would require the repeated use of the stochastic optimization tool developed in chapter 6.

Table 7-12. Earnings at risk – short-term risk for CHP system with various hedges

Short-term risk for October 2003	Short-term risk [T€]	Hedging quantities [GWh]		
		Electricity base	*Electricity peak*	*Coal*
Zero hedge	662			
Full hedge	555	778	0	1319
Peak/Off-Peak hedge	239	447	113	1170
Deterministic expected operation hedge	261	462	78	1277
Delta hedge	184	348	163	1239

Source: own calculations, based on prices available 1 October 2002

6.3.3   Integral earnings at risk

In Table 7-13, the integral earnings at risk obtained for the different power plant portfolios are summarized. Those do thereby not correspond to the sum of the delivery risk and the short-term risk computed for the same portfolio (same hedge). Instead the concept of integral earnings at risk is applied, defining the IEaR as a risk measure comprising all risks, which have to be borne by the operator but taking into account also his later risk hedging opportunities. Therefore to the short-term risk for a given portfolio the delivery risks obtained with a deterministic hedge are added. The deterministic hedge is chosen as the hedge for the delivery period since it is easily computed and corresponds to the current practice in the industry.

According to these results, a minimum risk of about 0.5 Mill. € occurs for all power plants. This corresponds to 10 % or more of the total expected earnings during this period. For the gas-fired plant, this risk is to a very large extent occurring during the delivery period, whereas for the coal-fired plant also the delivery price risk is quite important.

Table 7-13. Integral earnings at risk for power plants and CHP portfolios including various hedges

Integral earnings risk for October 2003 [T€]	Coal-fired plant	Gas-fired plant
Zero hedge	719	735
Full hedge	725	1091
Peak/Off-Peak hedge	725	602
Deterministic expected hedge	531	597
Stochastic expected hedge	556	598
Delta hedge	505	542

Source: own calculations, based on prices available 1 October 2002

Chapter 8

TECHNOLOGY ASSESSMENT
With application to fuel cells

A key uncertainty in the longer run is the development of technologies with respect to their efficiency, risks, environmental effects and costs. Technology assessment has become a growing field of its own (cf. e.g. Grunwald 2002, Braun 1998) and cannot be fully treated in the context of this study. Nevertheless, assessing the prospects and potentials of new power generation technologies is a key issue for power producers in order to develop strategic perspectives. And therefore a systematic approach is required to adequately cope with these uncertainties.

Such a methodology can much less than in the case of short and medium term decisions rely on a stochastic description of the uncertain factors. The decision problem is rather one of decision under uncertainty (in the narrow sense) than of decision under risk (cf. chapter 3). In order to reduce the uncertainty and to improve the decisions it is notably necessary to analyze and synthesize multiple informations coming often from various sources and being of very heterogeneous kind.

Clearly, technology assessment is in the electricity industry part of a broader activity sometimes labeled technology management. For example, Farrukh et al. (2000) distinguish the five steps of technology identification, technology selection, acquisition of selected technologies, exploitation of technologies and protection of knowledge within technology management. Since the major topic of the study is planning under uncertainty, the focus is laid in the following on the early stages of technology identification and technology selection. But in the electricity industry, the main subject of the technology identification step is an assessment of the prospects and potentials of known technology lines since the basic concepts of the new production technologies such as fuel cells, wind energy converters, PV cells

1. Analysis of the state of the art
2. Assessment of technology prospects
3. Assessment of cost developments
4. Assessment of context factors and resulting technology potentials
5. Strategy development

The methods to be applied within these steps depend often on the technology under study. Therefore, the approaches to be used are discussed along with an exemplary application for the case of stationary fuel cells. Fuel cells are chosen as an example since the past years have partly seen very enthusiastic prospects being propagated for this technology. It has been perceived as a unique technology which might revolutionize almost all energy relevant areas, be it in the transport, domestic or industrial sector. Additionally, fuel cells could be a first step from a more and more criticized fossil fuel based energy economy towards more sustainable energy supply structures. Whether these prospects are realistic will be discussed in the following sections, based notably on the work of Sander and Weber (2001) and Sander et al. (2003).

In general, the prospects of technologies may either be assessed based on technological research and considerations or on extrapolation using experience curves or simulation techniques. Of course the principal caveats on forecasting (cf. e.g. Grunwald 2002) apply to both approaches and therefore a combination of both seems most adequate. Therefore, the state of the art for fuel cells (section 1) and their technological development (section 2) are discussed based on a compilation and structuring of as much as possible available information. Then the concept of experience curves is applied in section 3 to assess the future economic prospects of fuel cells and a detailed simulation tool is used in section 4 to identify the impact of context factors and the resulting potentials for fuel cell applications. Finally, in section 5, strategic perspectives are discussed based on various, mostly informal models.

1.   ANALYSIS OF THE STATE OF THE ART

In the last years, fuel cells have completed huge steps towards a commercial use, notably in the stationary energy supply. Besides the phosphoric acid fuel cells (PAFC), which have been available commercially since 1992, also other technology lines have been tested successfully in full scale systems. Nevertheless, further significant improvements in efficiency and especially costs are necessary before commercial success can be expected. Therefore, examples of installed demonstration plants are first briefly discussed in this section, and then the future developments of technological and economic characteristics in the medium and longer term are pointed at in the following sections.

Currently, the following four technology lines are most considered for future commercial applications:

Polymer membrane fuel cells (PEFC) have a rather low operating temperature limiting the use of their waste heat mostly to room heating and hot water supply. Consequently the main developments are focusing here on small compact CHP systems for the residential sector (cf. Table 8-1).

An advantage for the PEFC is thereby the very rapid start-up and regulating performance. This offers the option to use the PEFC also for premium power production and grid backup in sensitive fields in the industry and service sector. In this second segment, with a modular power supply of up to 250 kW_{el}, notably the developments by Ballard Power Systems have to be mentioned. But even for these plants the option of waste heat use is foreseen in order to address further market segments.

Table 8-1. Demonstration plants for polymer membrane fuel cells (PEFC)

	Ballard Generation Systems Berlin-Treptow, D (6/2000+)	Vaillant Remscheid, D
Nameplate capacity	250 kW_{el} (factual: 212 kW_{el}) ca. 237 kW_{th}	4,5 kW_{el} 7 kW_{th}
Efficiency at nameplate capacity	Electrical: 35,2 % thermal: ca. 40 %	Electrical: ca. 35 % thermal: ca. 45 %
Temperature level useful heat	74 °C	60 °C

Phosphoric acid fuel cells (PAFC) have so far been produced mainly by International Fuel Cells (IFC), who are up to today the only ones to establish a fuel cell system on the commercial market (cf. Table 8-2). More than 200 plants of the PC25[TM]-type have been installed worldwide and since 1996 the third plant generation is distributed. Further research on PAFC systems is reported from Japan.

Table 8-2. Demonstration plants for phosphoric acid fuel cells (PAFC)

	International Fuel Cells (IFC) - PC25[TM] (about 200 plants worldwide since 1992)
Nameplate capacity	200 kW_{el} (0 – 100 % continuously variable) up to 220 kW_{th} at 60 °C (optional: 105 kW_{th} at 120 °C and 105 kW_{th} at 60 °C)
Efficiency at nameplate capacity	electrical: 39 – 42 % thermal: 38 – 45 % (depending on outlet temperature)
Temperature level useful heat	75 - 80 °C (max. 120 °C)

Molten carbonate fuel cells (MCFC) have considerably higher operating temperatures and consequently start-up and regulation are much slower. These cells are therefore less interesting for the field of space heating, but rather for the supply of process heat. The manufacturers therefore focus more on larger unit sizes which could also be coupled with turbines.

Two concepts may be distinguished here. On the one hand the external reforming of natural gas as pursued by M-C Power, on the other hand the internal reforming making use of the reaction energy and propagated by FCE and MTU (cf. Table 8-3). A particularity of the so-called „Hot-Module"-concept from MTU is thereby that all components operating at same temperature and pressure levels are integrated in one thermally isolated vessel in order to reduce the system costs.

Table 8-3. Demonstration plants for molten carbonate fuel cells (MCFC)

	Fuel Cell Energy (FCE) Santa Clara, USA (4/1996 – 3/1997)	MTU Friedrichshafen Bielefeld, D (7/2000+)
Nameplate capacity	2 MW$_{el}$ (factual: 1,93 MW$_{el}$)	300 kW$_{el}$ ca. 100 kW$_{th}$
Efficiency at nameplate capacity	electrical: 43.4 % - 43.6 %	electrical: 45 % thermal: ca. 23 %
Temperature level useful heat	No indications	450 °C

Among the *solid oxide fuel cells (SOFC)*, two development concepts have to be distinguished: the planar cell concept (Sulzer-HEXIS, ZTek and others) is comparable to the designs used for other fuel cell types, the tubular concept developed by Siemens-Westinghouse is on the contrary rather specific (cf. Table 8-4). One might even make further distinctions among the planar technologies between cells operating at the „conventional" SOFC-temperatures of about 1.000 °C and new developments which aim at a reduction of the operating temperature at about 700 to 800 °C.

Table 8-4. Demonstration plants for solid oxide fuel cells (SOFC)

	Siemens-Westinghouse		Sulzer-HEXIS Winterthur, CH
	Arnhem, NL (1/1998+)	Irvine, USA (12/2000+)	
Nameplate capacity	109 kW$_{el}$ 54 kW$_{th}$	220 kW$_{el}$	1,5 kW$_{el}$
Efficiency at nameplate capacity	electrical: 46 - 47 % thermal: 34 %	electrical: ca. 57 %	electrical: ca. 30 %
Temperature level useful heat	85 - 120 °C	No indication	300 °C

Also, SOFCs are developed for different application fields. On the one hand, the use of SOFCs in plants of medium and high capacity is envisaged for combined heat and power production or, in combination with gas turbines, for the pure electricity production. On the other hand, also small scale systems for the home energy supply are developed.

2. TECHNOLOGY DEVELOPMENTS IN THE MIDDLE AND LONGER TERM

The characteristics of a technology not yet on the market today are in general not easy to foresee. In the case of fuel cells, the technology characteristics of the first fuel cell systems to be put on the market can already be discerned today, but for the longer term developments no precise forecasts are possible, notably since the further technology development will be closely linked to the future market development and the financial resources generated through sales revenues. But the following development lines emerge from a multitude of analyses and company informations.

After a successful demonstration of the first systems and their introduction into the market, the producers will direct their resources primarily into an extension of the market potentials, as has been the case with the PC25™-systems from IFC. Simultaneously, the increasing operation experience will lead to stepwise system optimizations with the focus being laid on cost reductions. Moreover, increasingly larger local energy grids will be supplied through fuel cell systems by upscaling the different system concepts, with the objective to enlarge the market. Further developments will aim at the utilization of solid fuels and the development of multi-stage systems.

2.1 Upscaling and hybrid systems

Starting from modular plants for larger supply tasks an upscaling of the systems can be expected in the medium term (cf. Figure 8-1). Thereby, one may anticipate that the stack sizes developed so far will be used further and only the peripheral components will be upscaled and integrated.

In general, economies of scale lead to a decrease of the specific costs (per MW) with increasing plant size. In the case of conventional thermal power plants, this effect arises mainly due to the dominance of three-dimensional, volume-related processes (notably combustion, expansion etc.) where an upscaling leads to an under proportional increase in material requirements and costs. The fuel cell reaction is on the contrary a two-dimensional surface-related process. Therefore the economies of scale will mainly arise in the peripheral components, so that for the total fuel cell systems only limited economies of scale are to be expected.

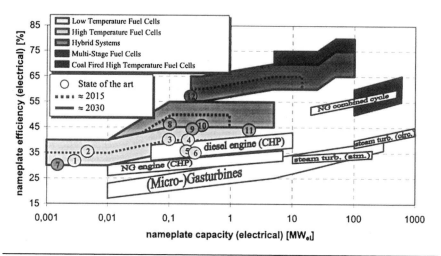

Figure 8-1. Unit size, efficiency and availability of different fuel cell systems and competing systems

Nr.	Manufacturer and technology		Electrical power [kW$_{el}$]	Electr. efficiency [%]
1	Hamburg Gas Consult	PEFC	3	32
2	Vaillant	PEFC	4,5	35
3	Fuji Electric	PAFC	100	40-40,2
4	International Fuel Cells (IFC)	PAFC	200	39-42
5	Mitsubishi Electric	PAFC	200	36
6	Ballard Generation Systems	PEFC	250	35,2
7	Sulzer-Hexis	SOFC	1	30
8	Siemens-Westinghouse	SOFC	100	46-47
9	MTU Friedrichshafen	MCFC	300	45
10	M-C Power	MCFC	250	44,4
11	FuelCellEnergy (MCFC)	MCFC	2.000	43,3-43,6
12	Siemens (SOFC – GT)	SOFC-GT	220	ca. 56

But the upscaling of systems may help to realize higher electrical efficiencies, notably through improvements in the following two fields:

- Efficiency improvements either for the peripheral components or through recuperation or reduction of parasitic energy flows (such as heat losses), which may only be done economically for larger plants or

- Realization of alternative system designs, which become only technically and economically feasible at larger plant sizes.

An important example of efficiency improvements for single components linked to increased plant size occurs in the case of hybrid SOFC systems. For those, electrical power outputs of up to 20 MW are expected until 2010. This leads, as indicated in Figure 8-2 to an increased efficiency of the gas turbine installed downstream of the fuel cell stack, raising the overall electrical efficiency by about five percentage points, under the (conservative) assumption of constant efficiencies for the fuel cell stack.

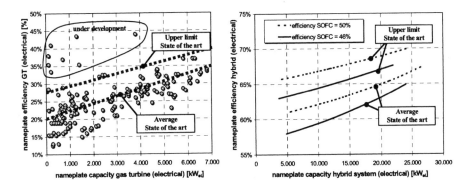

Source: Bohn et al. (2000) for gas turbines, own calculations for SOFC hybrid systems

Figure8-2. Efficiencies of gas turbines and SOFC hybrid systems depending on the unit size

The passage from increased efficiency for single components to new system designs is relevant in the case of gas turbines with intermediate reheating (cf. Figure 8-3). The multi-stage expansion can be realized here with conventional intermediate reheating or through appropriate configuration of two fuel cell modules corresponding to the two expansion steps of the gas turbine plant.

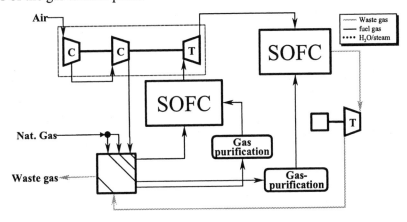

Figure 8-3. System design for SOFC with recuperative gas turbine and intermediate reheating

Similarly, topping an MCFC with a steam process becomes only attractive for larger plant sizes, as is the case for the additional integration of a steam process into SOFC hybrid systems (cf. Figure 8-4).

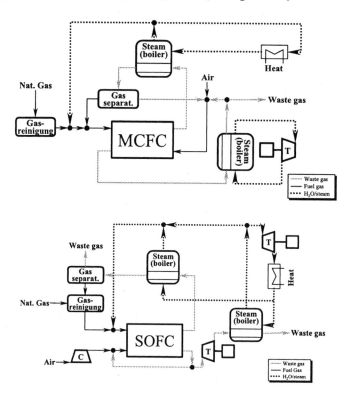

Figure 8-4. System designs for MCFC with steam turbine and SOFC with combined-cycle plant

2.2   Use of solid fuels

Fuel cell systems using coal or biomass as fuels only differ in the fuel processing substantially from the natural gas based systems discussed above. In principle, the systems have to be extended only at the boundary "gaseous hydrocarbons", prior to the reformer, including components for the thermal disintegration of solid fuels and purification of the raw gases obtained. Additionally the systems have then to be optimized to integrate optimally the heat flows including the new heat sources and sinks (cf. Figure 8-5).

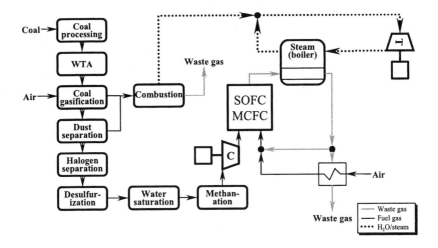

Figure 8-5. System design for fuel cells using coal as fuel

2.3 Multistage system concepts

For economical reasons, current plant designs restrict the fuel use in the fuel cell stack to 75 – 85 %, since the stack power output decreases with reduced fuel concentration in the combustion gas. The exhaust gases are usually burnt and the heat generated is used for the gas preparation or in a subsequent thermal power process. In view of an optimization of the systems for highest electrical efficiencies it is however desirable to use the gas as completely as possible within the fuel cell since even a downstream thermal power unit will never be as effective as the fuel cell reaction.

The power reduction of the fuel cell is due to the fact that the rear part of the stack, which is only poorly supplied with combustion gas, suffers from a reduction in electrical current intensity. Hence a possibility for increasing the fuel conversion rate consists in converting the fuel gas stepwise in a number of serially interconnected fuel cell stacks (cf. Figure 8-6). Every single stage can thereby work at its own voltage and amperage, so that a reduction in the rear cells does not lower the overall power output. At the same time, the fuel conversion rate has not to be maximized for each stage in a multistage system. A lower fuel conversion rate goes often along with a better working point for the stack, especially if the stages are optimized specifically to run with decreasing fuel gas concentrations and higher gas temperatures.

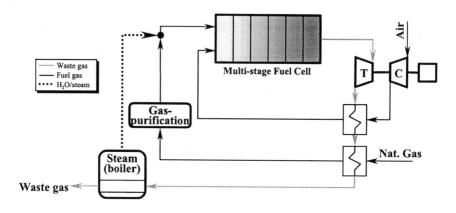

Figure 8-6. System design for multistage fuel cell systems

3.   COST DEVELOPMENT

The future costs for innovative technologies like fuel cells can either be assessed through bottom-up estimates, looking at the different components as in the previous section. Or they may be anticipated using the concept of learning or experience curves (cf. e.g. Wene et al. 2000) at a more aggregated level. Also both approaches may be combined, leading to an application of experience curves at the level of components or subsystems (cf. Ohl 2002, Sander et al. 2003). In the following the conceptual basis for learning curves is scrutinized first in section 3.1 and the use of experience curves for forecasting purposes is discussed (section 3.2). The dependency of experience curves on the stage of the technology development is analyzed in section 3.3. Existing data for stationary fuel cell systems are reviewed in section 3.4 and then an application at future developments for fuel cells is attempted in section 3.5.

3.1   Learning and experience curves – conceptual basis

With the increased use of a new technology, the users' knowledge on the application and the manufacturers' knowledge on the specific production processes grow continuously. This leads to a more and more adapted use and a continuously improved production of the technology and consequently to increased efficiency and lower production costs.

The phenomenon of continuous cost reductions in the manufacturing of products was first observed by Wright (1936), who described a causal relationship between the production costs of a product and its cumulative production quantity for the example of airframes in the aeronautic industry. Wright found that every doubling of the cumulative production quantity induced a decrease in the total working hours necessary by an almost constant percentage, and attributed that to „improvements in proficiency of a workman with practice" (Wright 1936, p. 123), emphasizing hence the increased physical and mental skills of the workers through experience.

With the conduction of numerous studies (e.g. Argote and Epple 1990, BCG 1972, Dino 1985, Dutton and Thomas 1984, Ghemawat 1985, Yelle 1983) the interpretation of the underlying learning processes shifted away from the original "learning by doing" (cf. Arrow 1962), in that the cost reduction could be attributed to more far reaching reasons such as new or optimized production processes, improved methods, substitution effects, production oriented redesign and standardization. (Horlock 1995, Claeson 2000). Hence many experience curves show typical discontinuities („knees"), where the cost reduction is too strong to be solely attributed to improved personal skills (cf. Figure 8-7). Rather these knees may often be attributed to technology leaps in the manufacturing process or in the technology itself, as illustrated in Figure 8-7 for the case of thin film photovoltaic cells (cf. Wene (2000)).

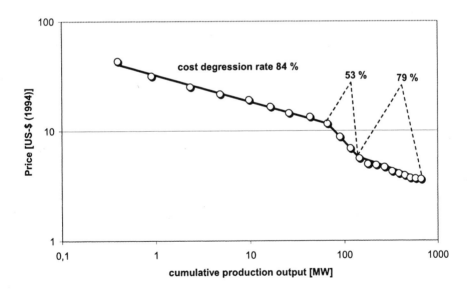

Source: Wene (2000), data taken from Goldemberg (1996), Nitsch (1998)

Figure 8-7. Discontinuities in experience curves as a consequence of technology leaps

Experience curves thus describe the cost development for a product starting from the first piece until the n-th copy produced, accounting for all relevant influences. However it has to be retained that the concept is mostly empirically motivated, and particularly no theoretical underpinning exists which could explain the specific shape of the cost reduction curve and/or its slope (Wene 2000). Experience curves are hence a macroscopic model, based on the ex-post analysis of empirical data (Tsuchiya 2000). A strong point of the approach is however its solid empirical underpinning, with a multitude of empirical observations documenting the learning effects with increasing production unambiguously. Notably, comparative studies clearly show that the learning rates vary considerably between different industrial sectors, nevertheless typically values between 70 % and 85 % are observed (cf. Figure 8-8). This makes it possible to use the empirical observations as guidelines for assessing the possible cost development for new technologies.

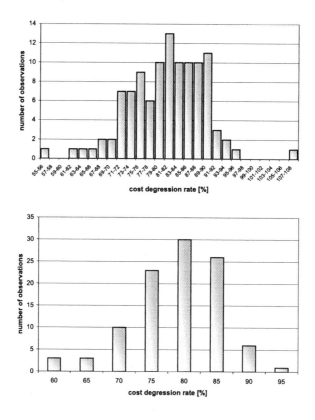

Source: Dutton and Thomas (1984), Ghemawat (1985)

Figure 8-8. Distribution of learning rates in different empirical studies

3.2   Experience curves as a forecasting tool

The hugest difficulty in the use of experience curves for ex-ante analyses arises from the fact that an exact prediction of the cost reduction curve is not possible. Therefore, both for the starting point as for the cost degression rate assumptions have to be made. One possibility is to investigate different cases, like this has been done by Lipman and Sperling (2000) in an analysis of the cost development for PEFC modules for automotive applications (cf. Figure 8-9).

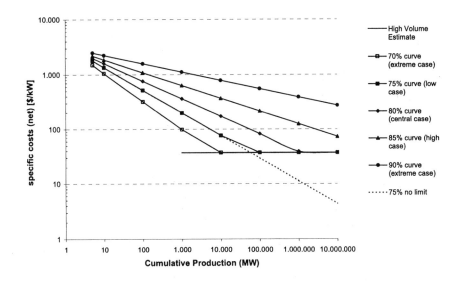

Source: Lipman and Sperling (2000)

Figure 8-9. Projections for the development of PEFC module costs for automotive applications including best- and worst-case scenarios

Such a best- and worst-case analysis leads however to a very broad range of results if a longer period is considered. In the analyzed case, the cumulative production quantity necessary to reach 37 $/kW$_{el}$ varies by a factor 1,000. Hence a narrowing of the projection interval is desirable, yet this requires the use of additional information sources.

Ex-post-analyses of similar products are a possible source of information. Furthermore cost data on executed demonstration or pilot plants help to determine the starting points. Yet this requires an extension of the cost reduction curves into the field of pre-commercial plants.

3.3   Dependency of the experience curves on the development stage

The introduction of new technologies usually follows the phases of:
1. research,
2. development and demonstration,
3. commercialization and
4. market penetration.

Especially if the existing technologies already fulfill the needs of the market, as is the case in the electricity production, a new technology has therefore already to go through several stages to achieve advantages compared to existing systems before being introduced on the market.

The first development stage is characterized by fundamental research on small and smallest systems followed by a gradual upscaling of the technology towards commercially attractive sizes. The learning process is rather difficult to describe for this stage, but when the subsequent development and demonstration phase is entered, the learning process starts on the real-scale technology, which may be described by an experience curve.

Nakicenovic (1997) shows for the example of the introduction of gas turbines that the learning process is rather different in the various development phases (cf. Figure 8-10). Namely the demonstration phase is characterized by a very dynamic development at the product level, with first "real" operation experiences resulting in rapid progress. Furthermore only at this stage a targeted adaptation of peripheral components is possible, since the functionality of the system is the key focus before the first demonstration, whereas manufacturing, cost considerations and the appropriate choice of complimentary are only getting important at later stages. For stationary fuel cells, this is particularly true for the "balance of plant", which consists mostly of existing technologies, which however have to be adapted to the specific requirements of the fuel cells.

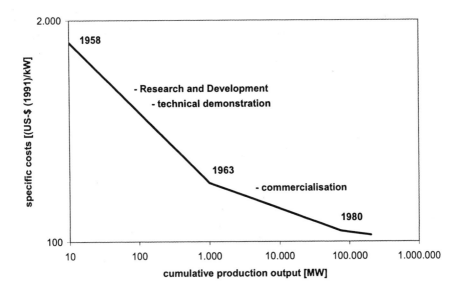

Source: Nakicenovic (1997)

Figure 8-10. Cost reduction curves for gas turbines accounting for different development phases

Figure 8-10 as well as the results of other studies indicate, that the learning capabilities are always reduced to a certain extent when the technology passes from the development and demonstration stage to the actual production (cf. also Kydes 2000). In the case of gas turbines, the corresponding cost reduction curve passes from an 18 % slope to a value of about 7 % during the commercialization phase. Certainly, one reason is that over-proportional cost reductions through exhaustive redesigns of the product are not the key priority after the targets for market entrance have been reached. Rather requirements on availability and reliable operation are then at the forefront.

3.4   Ex-post observations of cost reductions for stationary fuel cell systems

Hints at the cost reduction rates for stationary fuel cell systems during the commercialization phase can be obtained from the experience curve of the so far only commercially available plant, the PAFC system PC25™ which is sold since 1992 by IFC. Figure 8-11 shows the specific production costs for the system until 1999 as well as the experience curve at a 75 % learning rate (solid line), as proposed by Whitaker (1998). Stationary systems based on other technology lines differ in details substantially from the PAFC, but the overall system design comprising cell stack and balance of plant is rather similar. Therefore, a learning rate of 75 % in the commercialization phase seems in general adequate for stationary fuel cell systems.

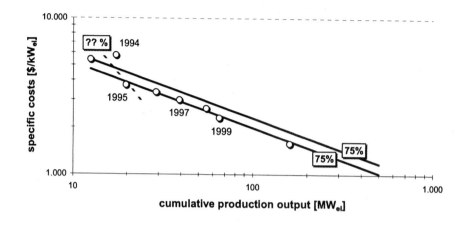

Source: Whitaker (1998), Sander (2001)

Figure 8-11. Cost reduction curves for the PC25™ systems from IFC

Whitaker (1998) takes as starting point for the experience curve shown in Figure 8-11 a value of about 5.500 $/kW$_{el}$ at a cumulative production quantity of about 12 MW$_{el}$. From the diagram it is however evident that a downward shift of the experience curve considerably improves the fit to the data points given from 1995 onwards (lower line). In line with the considerations of the previous section, this points at a considerably steeper cost reduction during the demonstration phase, even if the available data are not sufficient for quantification.

For gas turbines, the cost reduction in the demonstration phase was almost the double of the one obtained in the commercialization phase. This would imply a learning rate of 50 % for fuel cells. However, this would correspond to a cost development which is typical for a technology leap. Obviously, fuel cells are a completely new technology with correspondingly high learning potential, but the balance of plant is mostly based on conventional components. For those, specific adaptations are necessary, over-proportional learning effects may however not be expected.

An approximate calculation of the learning effects during the demonstration phase may hence be done by maintaining for the balance of plant the learning rate of 75 % as observed for the PC25™-systems, but assuming a technology leap for the stack itself with about 50 % cost degression rate. Data from Keitel (2000) indicate that about half of the system costs are due to the balance of plant, so that learning rate for the overall system during the demonstration phase would be around 60 – 65 %.

Thus, it is now possible to indicate a range for future cost reductions based on economic data for the first demonstration plants and to compare these to the plans for market introduction.

3.5 Projection of experience curves for different technology lines

Economic data for the MCFC technology can be obtained from the first demonstration of the system design by the American company FuelCellEnergy. This plant started operation in 1996 in Santa Clara, California and had investment costs of about 20,000 $/kW (FCCG 2001). The further points on the upper line in Figure 8-12 correspond to the plant installed in 1999 in Danbury and the further five pilot plants, originally scheduled until 2003.

Figure 8-12 comprises also the first data for the "Hot-Module"-MCFC-design by the German manufacturer MTU Friedrichshafen with costs of 10,000 $/kW$_{el}$ (University Bielefeld 2001). The further data points correspond thereby to the plant installed in 2001 in Bad-Neustadt, the further pilot installations foreseen in the next one or two years and the market introduction which is planned thereafter.

For the cost development of systems based on the polymer membrane fuel cell (cf. Figure 8-13), a starting point with about 15,000 $/ kW$_{el}$ can be derived from the 250 kW$_{el}$-demonstration plant in Berlin-Treptow (Keitel 2000). The start of the commercialization has been scheduled here with the end of the planned zero series around 2003.

Somewhat more difficult is an estimation for the systems based on the SOFC technology. Demonstration plants have been put in place by Siemens-Westinghouse, but two plant designs have been developed in parallel.

Cost figures are available for the CHP variant, which has been put into operation in Arnhem in 1998 (about 100,000 $/kW$_{el}$, Kuipers 1998). The second concept is a combination with gas turbines, where a first pilot plant has been put into service in Irvine (USA) in the year 2000. The experience curve shown in Figure 8-14 presumes that these hybrid systems also contribute to the learning effects for the CHP variant and vice versa.

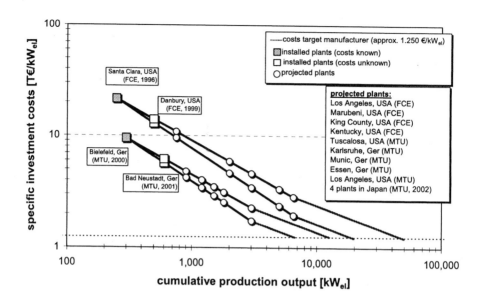

Source: Sander et al. (2003)

Figure 8-12. Estimation of cost reduction curves for MCFC systems by FuelCellEnergy (FCE) and MTU Friedrichshafen („Hot-Module")

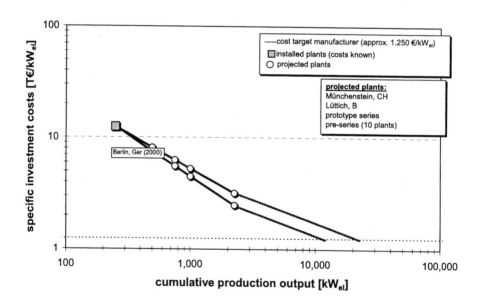

Source: Sander et al. (2003)

Figure 8-13. Estimation of cost reduction curves for the PEFC-System from Ballard Generation Systems

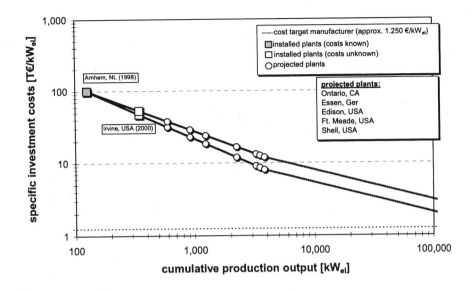

Source: Sander et al. (2003)

Figure 8-14. Estimation of cost reduction curves for the SOFC system from Siemens-Westinghouse

Given the theoretical difficulties and the limited empirical data base, the cost reduction curves shown can certainly only be considered as approximate estimations. Yet they clearly indicate that important cost reduction and development efforts are necessary before market penetration can be achieved. However, the on-going efforts clearly point in the right direction. What remains to be seen is what effects technology specific synergies and context factors will exert. For example, a close cooperation exists in the case of the MCFC between the two manufacturers. The PEFC could on the other side benefit from the developments for mobile applications. Finally, the SOFC is still at a rather early development stage with pre-commercial system sizes, so that possibly a far higher cost reduction could be achieved.

4. ASSESSMENT OF CONTEXT FACTORS AND RESULTING TECHNOLOGY POTENTIALS

For many new electricity-producing technologies, the application is strongly dependent on site-specific factors. In the case of renewables, it is the availability of wind, sun or biomass; in the case of fuel cells, micro gas turbines and Stirling engines, it is the existence and load shape of an appropriate heat demand.

The methodology for the assessment of potentials consequently depends on the type of technology considered. Here, the focus is on stationary fuel cell applications, where application potentials in industry and the residential sector are scrutinized using detailed demand and load shape data. The analysis is carried out for the state of Baden-Württemberg, a region with about 10 Mill. inhabitants in the south-west of Germany.

4.1   Energy demand structure

A prerequisite for a realistic assessment of the potentials and the market development for stationary fuel cells and other CHP systems is a detailed data basis on the object specific energy demands and load curves. This has to be worked out on the basis of official statistics and the detailed analysis of representative concrete objects. Specifically for the question under study, the energy balance of Baden-Württemberg (cf. Wirtschaftsministerium Baden-Württemberg 2002), disaggregated by energy carriers and demand sectors, can be taken as starting point.

However, this disaggregation is far from being sufficient. In the case of private households, the energy demand especially depends on the following factors (cf. Ebel et al. 1990, VDEW 1993, Weber 1999):

Heating energy:
- Building size
- Building type (single family house, large dwelling block etc.)
- Building vintage

Tap warm water use:
- Number of persons in the household

Electricity use:
- Equipment with household appliances
- Number of persons in the household
- Use of air-conditioning and ventilation

Of course, further factors are important, but these are either hardly observable (such as behavioral factors) or they are variable in the context of the decision problem under study (heating energy carrier and heating system).

For the analyzed case, the energy use for heating is most important, since it accounts for about 75 % of all residential energy use in Germany and Baden-Württemberg (AG Energiebilanzen 2001, Wirtschaftsministerium Baden-Württemberg 2002). Based on earlier studies (notably Blesl 2002, Kolmetz, Rouvel 1995, VDEW 1995, Berthold et al. 1999, Blesl and Fahl 2002), the energy demand structure as shown in Figure 8-15 can be derived.

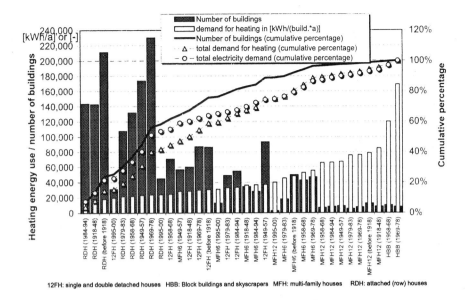

Source: Sander et al. (2003)

Figure 8-15. Final energy use of households in Baden-Württemberg

The buildings are here grouped by types and vintages and then ordered by increasing energy demand for space heating. About 80 % of all buildings have a heating energy use of less than 30,000 kWh per year. Furthermore, one observes a rather uniform distribution over the building types and vintages. Notably, the 20 % of all buildings with the lowest space heating energy demand per building (the lower 20 % quantile) still account for more than 10 % of the energy use in that segment. Conversely, the upper 20 % quantile consumes about 35 % of the total energy use.

For the energy demand in the industrial sector, besides the energy uses mentioned so far also process heat has to be accounted for. Furthermore the company size, as measured by the number of employees is an important factor for explaining the total demand. The problem is further complicated by the fact that energy cost data are compiled and published in official statistics differentiated by company size categories. But physical energy demand is not available and the link between costs and quantities is non-linear, due to degressive tariff structures. But through a newly developed methodology, it is possible to attribute the total energy use with the best possible fit on the different industry sectors, company sizes, energy carriers and energy uses. Figure 8-16 shows the resulting energy use disaggregated by industry sector and company size category.

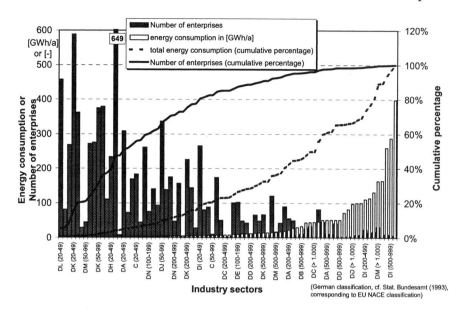

Source: Sander et al. (2003)

Figure 8-16. Final energy use of industry in Baden-Württemberg

It turns out that small- and medium-size enterprises clearly dominate in Baden-Württemberg and that consequently far more than 90 % of all companies have an energy use of less than 50 GWh per year. By contrast, a few large companies (including e.g. DaimlerChrysler and Bosch), which represent only about 3 % of all firms, cover almost half of the total consumption. The upper 20 % quantile here accounts for 80 % of the total industrial energy use, the lower 20 % quantile on the contrary less than 1 %.

Already at this stage it hence becomes evident that small units are advantageous for reaching large market potentials both in industry and in the residential sector. To what extent the currently available system sizes restrain the usage possibilities, is analyzed in the following based on detailed operation simulations.

4.2 Operation simulation

In order to identify advantageous market segments for fuel cells, it is necessary to assess their relative profitability compared to already established alternatives. This is true since stationary fuel cells do not provide energy to so far unsupplied application cases, but compete with other technologies, notably in the field of decentralized CHP. For comparing the performance and the costs it is necessary to make detailed simulation studies which account for the characteristics of the different application cases, such as the heat and the electricity demand (total annual values as well as load profiles), the temperature level of the heat and the simultaneity of electrical and thermal load. These parameters will influence the profitability of fuel cell applications especially since purchased electricity and electricity sold to the grid are currently - and will also be mostly likely in the future - valued differently. Therefore a detailed analysis of demand data including the identification of object-specific load profiles is of primordial importance, using e.g. the approaches developed by Weber et al. (2001).

4.2.1 Principles of operation simulation

Once the demand data have been identified, an operation simulation for the alternative supply options has to be carried out[55]. This should be done with high temporal resolution in order to be able to account for part-load and dynamic characteristics of the competing options. From these simulations the annual system costs can be identified as a function of system design parameters such as the share in maximum load covered by the fuel cell or CHP system. Thereby capital costs are annualized using an annuity factor corresponding e.g. to 20 years lifetime. If energy prices or other relevant parameters are expected to vary also substantially during the lifetime of the supply unit, a discounted average of the expected prices should be used to avoid carrying out simulations for multiple years.

As a result of such simulations, the net present value (or rather cost) of different supply options is obtained as a function of the size of the CHP unit as shown in Figure 8-17 for large enterprises in the chemical industry. The first series of simulations aims thereby at identifying the benchmark for fuel cell systems – which is in the case of the large chemical firms a gas-fired motor-CHP unit with 4,800 kW thermal nameplate capacity. For this configuration, the lowest net present cost is observed in Figure 8-17. In many other applications cases, the lowest costs are however obtained for the electricity purchase plus boiler option. Conventional CHP options in general are close to the economic viability, but only in a few large enterprises they are clearly advantageous.

---

[55] Thereby it is assumed that the system operation can be described by simple decision rules. If the system has on the contrary several degrees of freedom, it might be necessary to run optimizations similar to those described in section 6 to determine the optimal system operation.

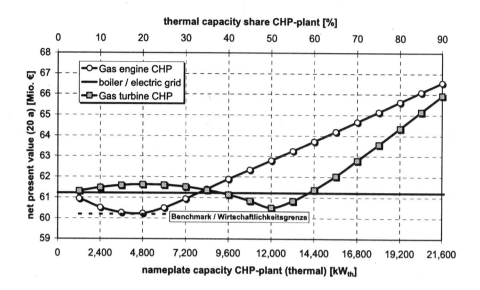

Source: Sander et al. (2003)

Figure 8-17. Net present value of conventional technologies in large enterprises (more than 1,000 employees) of the chemical industry and determination of the economic benchmark for fuel cells

In these cases, about 20 % to 50 % of the maximum thermal load is typically covered by the optimally designed CHP unit. The precise share depends especially on the electricity to heat ratio of the chosen CHP system.

As for conventional CHP units, the power of stationary fuel cells has to be adapted to the electric and thermal load profile of the analyzed segment to obtain an economically optimal operation. This design optimization is done within the operation simulation iteratively. At given investment costs, the net present costs of all possible system sizes are determined first, increasing gradually the fuel cell output power up to the maximum demand in the segment considered. Simultaneously the output power of the peak load boiler is adapted to the remaining demand. The optimal plant design then corresponds to the size with the lowest net present costs. If these costs exceed for all system sizes the net present costs of the benchmark system, the investment costs are gradually reduced, until the benchmark is reached for the first time. This approach is illustrated in Figure 8-18, where the net present costs for high temperature fuel cells are plotted against the thermal output power of the fuel cell unit.

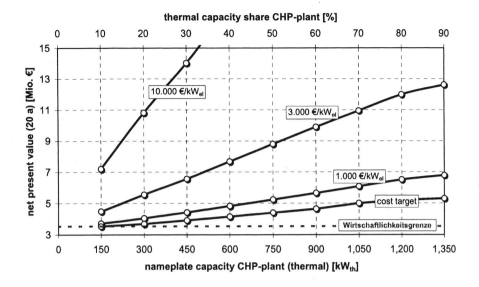

Source: Sander et al. (2003)

Figure 8-18. Iterative determination of target costs for the example of high temperature fuel cell systems in small enterprises (50 to 99 employees) of the chemical industry

From these calculations it turns out that fuel cells are much smaller than conventional CHP if optimally designed, with a share of less than 10 % in maximum thermal load (cf. Figure 8-19). The optimal size usually does not differ substantially between high temperature (MCFC, SOFC) and low temperature (PEFC) systems. Only hybrid systems exhibit somewhat smaller optimal sizes, a consequence of the higher electricity to heat ratio compared to low temperature systems. This low share in thermal capacity is in many segments below the single module size of about 250 kW_{el} for the envisaged early market products.

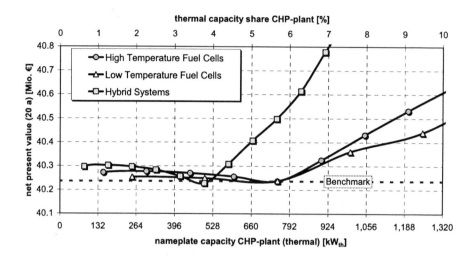

Source: Sander et al. (2003)

Figure 8-19. Net present value of fuel cell systems in large enterprises (more than 1,000 employees) of the metal industry

For the shape of the net cost curves, two basic alternatives may be observed. Typically, the pattern is similar to the one observed for gas turbines (cf. Figure 8-17) with an increase in costs for small units followed by a minimum and a subsequent strong increase. For low-temperature systems, the pattern obtained is often comparable to the one for motor CHP units, which is not surprising given the similar characteristics of these two technologies (cf. Figure 8-20).

The strong cost increase at unit sizes beyond the optimum is mostly occurring when the customer's own electricity demand is almost covered and the additional electricity generated is mainly fed back into the public distribution grid. But with increasing company size, sometimes a second cost minimum is observed for high temperature and hybrid systems, which indicates that in these cases also excess power can be sold profitably (cf. Figure 8-20).

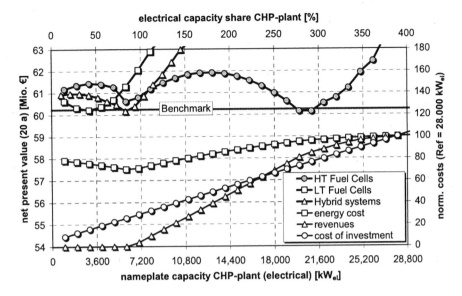

Source: Sander et al. (2003)

Figure 8-20. Net present value of fuel cell systems and surplus electricity generation in large enterprises (more than 1,000 employees) of the chemical industry

Figure 8-20 also shows the relative development of the key cost categories for the high temperature systems. Obviously, the decrease in purchase costs (fuel and electricity) leads to the initial minimum. Thereafter, additional sales revenues do first not compensate the increase in purchase cost. But when more electricity is fed in, the price paid is increasing, given the increasing number of full load hours provided. This then leads to the formation of the second observed cost minimum.

Therefore a substantial excess power production potential is observed in selected sectors, (cf. Table 8-5), which could be interesting for the electricity purchase of utilities or the market entrance of new independent power producers (IPPs).

Table 8-5. Industrial sectors with potentials for excess power production when target costs are reached

| Sector | Excess potentials (Output FC/ Maximum load in [%]) | | | |
| | High temperature systems | | Hybrid systems | |
	Power	El. production	Power	El. production
Alimentation	300	360	150	300
Pulp and Paper	170	190	190	160
Chemicals	290	280	-	-

Source: Sander et al. (2003)

All results on target costs and excess power potentials are obviously strongly dependent on the future price relations in the market. The calculations shown here have taken the current German association agreement (Verbändevereinbarung IIplus, cf. BDI et al. 2001) as basis for the grid fare calculations and the models described in chapter 9 for the calculation of future generation costs.

4.2.2   Aggregated utilization potentials in Baden-Württemberg

From the simulations of the different market segments considered, the graphs in Figure 8-21 can be derived, describing the required target costs as a function of the total final energy demand. In the pessimistic scenario presented in Figure 8-21, the target costs are in many segments around 500 to 600 €/kW$_{el}$.

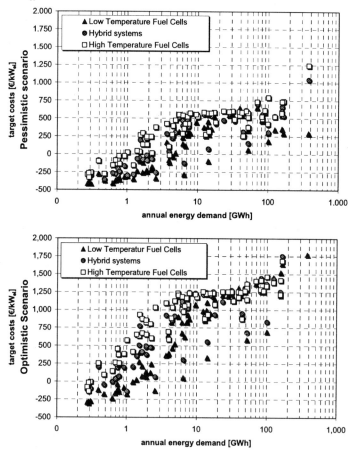

Source: Sander et al. (2003)

Figure 8-21. Target cost to be achieved for fuel cells in different industrial sectors

Only for some large customers, a few cases of considerably higher target costs are observed. Under higher electricity and lower gas prices, the target cost get considerably higher, reaching more than 1,000 €/kW in many application cases.

For smaller application cases, a net reduction in the target costs to be achieved is observed, which results from oversizing of the envisaged pioneer market products with minimal power output of about 250 kW$_{el}$. Low temperature systems are in general somewhat disadvantaged. Here oversizing occurs earlier due to a lower electricity to heat ratio and the limitation of heat use to low-temperature applications. The role of heat temperature levels is also evident for the few points with low target costs at high energy demand. Those appear for sectors, where process heat demand at low temperature levels is low or inexistent, notably the production of rubber and plastics.

These analyses of single application cases can now be aggregated to obtain the economically viable potential of stationary fuel cells as a function of the achieved target costs (cf. Figure 8-22).

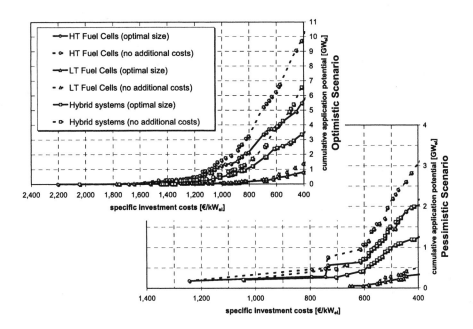

Source: Sander et al. (2003)

Figure 8-22. Economic potential of stationary fuel cells in the industry of Baden-Württemberg as a function of specific investment costs of the fuel cell systems

Thereby it has to be taken into account that in those sectors, where profitability has already been reached, the potentials are also rising with diminishing costs. Furthermore, the distinction has to be made between the cost minimal potentials and those (larger) potentials, which are realizable without extra costs compared to conventional systems.

For installations in the residential sector, the first economic potentials already appear at considerably higher costs (cf. Figure 8-23). Already above 2,000 €/kW first potentials appear, if the current grid tariffs are also applied to households with own generation. In these grid tariffs, the grid costs, which are mostly capacity dependent, are charged proportionally to the energy consumed – assuming that energy consumed and peak load are rather proportional. This relationship is certainly different in the case of households with own fuel cell generation. Therefore a second set of scenarios has been computed where the households have revenues (or saved costs) from own generation corresponding to the current subsidy to fuel cell system based generation of about 5 €ct/kWh. In all cases, not only the costs of the fuel cell system itself are included but also the investment into the necessary peak load burner.

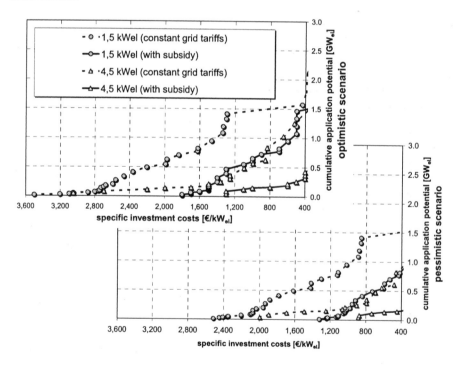

Source: Sander et al. (2003)

Figure 8-23. Economic potential of stationary fuel cells in the residential sector of Baden-Württemberg as a function of specific investment costs of the fuel cell systems

4.2.3   Sensitivity analyses

Sensitivity analyses provide important insights into the relevance of different parameters for the overall potential assessment, even if they further widen the range of results obtained. In the case of fuel cells, we have notably analyzed the impact of reduced variable operation costs, which result from a doubling of the stack life time.

This has a rather limited impact (cf. Figure 8-24), since especially oversized systems do hardly benefit from such a development. For larger application cases, the target costs however increase by about 20 %, for the largest cases even up to 50 % are possible. The improvements are relatively much higher for low temperature systems with 75 % up to 200 % improvement in particular cases.

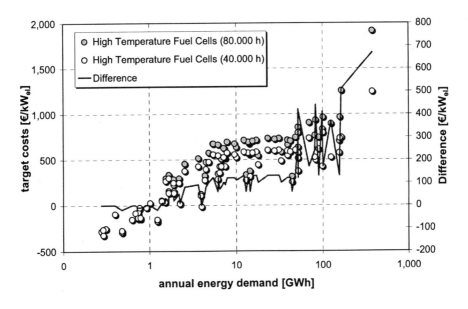

Source: Sander et al. (2003)

Figure 8-24. Impact of reduced operation costs through doubled stack life time on the target costs of fuel cell systems in industry

As mentioned already, the envisaged module size leads to a reduction of target costs in the field of small and medium sized enterprises. This means that target costs should rise in all those cases, where the system has been oversized so far. Additionally the somewhat higher electricity prices paid by small customers increase the profitability. In Figure 8-25 it is shown that this leads in certain cases to target costs which are even higher than those for larger objects. Such a smaller module size could be especially beneficial for low temperature systems, since they may also become attractive in sectors with reduced low-temperature heat demand.

However, one has to keep in mind the economies of scales related to module size. As shown in Sander et al. (2003), one would expect investment costs for small-scale systems with 25 kW$_{el}$ to be about a factor two higher than those for 250 kW$_{el}$ modules. Consequently, the shown improvements in target costs may not be attractive enough for moving rapidly towards these small scale systems.

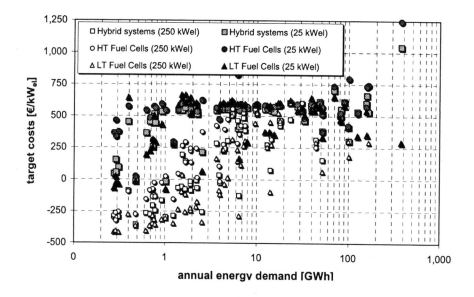

Source: Sander et al. (2003)

Figure 8-25. Impact of reduced module size on the target costs of fuel cell systems in industry

5.   IDENTIFYING TECHNOLOGY STRATEGIES

Despite the qualitative and quantitative assessments carried out in the previous sections, the situation for utilities considering investment in fuel cell technologies remains one of decision making under uncertainty of type III as distinguished by Courtney et al. (1997, 2001) (cf. chapter 3, section 3.1). There is a true uncertainty on the future costs of fuel cell systems, which cannot right away be reduced to a few discrete cases. On the contrary, the costs in the longer term may vary on a continuous range from values that limit fuel cell use to some selected niche markets up to a range where fuel cells are effectively competitive in a wide field of applications in households and industry. Cost uncertainties arise both from the technology itself and its production costs as from the context factors including notably gas prices and potential CO_2 costs. Obviously there is also little chance to assign objective probabilities to the different cost scenarios in the current situation, especially since technology development is truly uncertain and also the application of probabilities to political decisions would be rather speculative. Nevertheless utilities have to chose right now a strategy how to cope with this emerging technology.

This strategy does not need to be fixed forever but rather should be flexible and include clear targets and milestones – as shown empirically by Grant (2003) for the strategy of the oil majors. What are then the required actions, in the terminology of Courtney et al. (1997): doing "big bets", realizing options or taking no-regret measures? Courtney et al. (1997) emphasize that this depends on the strategic posture taken, distinguishing the postures of shaping, adapting and reserving the right to play. But at the current stage, big bets in the form of huge investments in fuel cell plants seem clearly not adequate, since this would be prohibitively expensive at current plant costs (cf. section 3). A possible big bet could also be vertical integration with a fuel cell supplier. But as Sutcliffe and Zaheer (1998) show, vertical integration is not a preferred action in situations with considerable primary uncertainty – and technology uncertainty is one of the possible sources of primary uncertainty mentioned by them. The risk of vertical integration in this case obviously is to make the false bet and to be integrated with a company which is not able to deliver the promised resource (technology).

On the other extreme of possible actions, so far hardly any no-regret measures on fuel cells can be identified, except perhaps the collection of relevant information. An engagement in demonstration plants already requires considerable resources and is not sure to be rewarded later on. Therefore, it should rather be seen as realizing an option. An option, which has partly a pay-off in the future by having increased the capabilities of the employees and the organization to handle such an innovative technology (cf. Grant 2003). But part of the pay-off is also in the present time. Given the positive public image associated with innovation in general and with fuel cells in particular, a demonstration activity is a possibility for transforming the public image of a utility and of getting media citations at low costs compared to advertising campaigns. However, two points have to be noted here: on the one hand, a transformation of the public image is probably most effective in the region where the demonstration plant is implemented. Therefore, this argument is particularly valid for smaller regional and municipal utilities. On the other hand, the image effect is in all likeliness decreasing over time – to the extent that the general public gets used to information on successful fuel cell demonstrations, it will become less attentive to such media messages. Also the effect of capability building is questionable in the case of a technology, which will take at least five further years for penetrating the market on a substantial scale.

From these potential developments, a scenario could arise that despite good long-term prospects, utilities are not willing to invest in the fuel cell technology in the medium term since it is neither economically attractive nor appealing from a public relations point of view. If political support is also lowering – e.g. given severe budget constraints – it could well happen that the technology is not pursued further despite a potential win-win situation in the long run, creating a scenario of technology lock-in as arising frequently in the case of increasing returns to scale, cf. e.g. the discussion in Mattson and Wene (1997) or Cowan (2000).

But also other scenarios are conceivable. E. g., RWE has committed itself publicly in 2001 to a strategy with high shares of decentralized generation including fuel cells in the longer term (Jopp 2001). It has then clearly adopted a strategy of shaping, putting forward ambitious goals for the penetration of fuel cells up to the year 2010. Other utilities have been less affirmative, but engagements in pilot and demonstration activities such as those reported by EnBW (2002a) clearly go beyond a pure adaptation strategy. It is a strategy of realizing options and if these options are not all exercised[56] this does not necessarily mean that the technology line will not be pursued further.

---

[56] EnBW (2002b) recently announced that it will not realize a 1 MW SOFC plant planned together with Siemens-Westinghouse. But they maintain their demonstration activities with Sulzer-Hexis (small scale SOFC) and MTU (MCFC).

Chapter 9

INVESTMENT DECISIONS

Until 2030, the requirements for replacement investments in the EU power plant sector are expected to be between 300 and 600 GW (cf. Voß 2002, Birol 2003). Also elsewhere in the world the generating capacities built up during the high growth phase in the 1960s and 1970s will have to be replaced and new capacities have to be installed, leading to investment requirements of about 5,000 GW between 2000 and 2030 (cf. IEA 2002, Birol 2003). This requires strategic decisions for electric utilities since power plants require large investment sums and represent, once effectuated, to a large extent sunk costs[57].

Since electricity demand has been growing steadily albeit slowly in the last decades, there is hardly any uncertainty on the necessity of building up new capacities – although in the first years after the liberalization of European electricity markets only few people considered the issue of future power plant investments (cf. e. g. Voß 2000, Voß 2001, Weber, Voß 2001, Pfaffenberger 2002). But a first uncertainty in this field is whether political measures will lead to a situation where all new generation capacities are supplied by subsidized or regulated segments within the electricity sector - notably generation from renewable sources and/or generation from combined heat and power including fuel cells. In Germany, this possibility is sometimes evoked. But overall, such a scenario seems not very likely, given that fuel cells will certainly not be fully competitive in the next decade, other combined heat and power generation is currently only competitive in selected applications and wind – as the most important renewable source – requires always conventional back-up capacities in order to cope with the

---

[57] At least in continental Europe so far hardly any market for second-hand power plants exists – in the UK and the US sales of used power plants have been more frequent in the context of divestment strategies.

intermittent character of wind generation. A further alternative, mentioned sometimes, is the increased use of electricity imports to avoid new construction in one specific country. But this is certainly not a solution for Europe as a whole and even for a larger country like Germany, recent analyses indicate that the potential for imports is limited through several factors: current grid constraints between countries, the requirements of system stability and limitation of transportation losses and the expectation that in the longer run technology and fuel costs will tend to level out within Europe, thus not creating any large incentives for installing power plants far from the point of use. (cf. Fichtner, Cremer 2003). Overall, the general uncertainty on the need for new conventional plants is rather low and obviously these will be mostly thermal power plants, since hydro potentials have been already exploited to a large extent in the past (cf. Witt, Kaltschmitt 2003).

Nevertheless, there is considerable uncertainty on what power plants will be most cost efficient over the considered lifetime. This is partly due to uncertainties of primary energy prices, which can be modeled as stochastic processes (cf. chapter 4, section 3.1). Notably with the raise of gas prices since mid 1999, it has become rather questionable whether gas-fuelled power plants are *the* option of choice for new power stations. Coal, nuclear and (where available) lignite are potential alternatives.

But for deciding which one to chose, other uncertainties have to be taken into account and notably uncertainties related to political processes, to which no objective probabilities can be assigned. Hence as in the case of fuel cells, there is a level III uncertainty in the terminology of Courtney et al. (1997). Especially, a continuous range of values is possible in the case of climate change levies or Greenhouse gas certificate prices. Other political uncertainties, notably those surrounding the future use of nuclear energy in a number of countries in Europe and worldwide, have to be added, but these may be rather viewed as discrete scenario Type II uncertainties.

In this context, novel methods are needed to assess the profitability under uncertainty of the various types of thermal power plants. In principle the application of a real options approach (as notably developed by Dixit and Pindyck 1994) seems desirable, but it is complicated by the possibilities of fuel switching with endogenous output prices and simultaneous uncertainties on loads, fuel prices, technologies and policies. A simple extrapolation of current price patterns to the future seems problematic since current price patterns, as analyzed in chapter 4, hardly reflect capacity constraints, given that in the past substantial overcapacities existed in continental Europe. But this situation will come to an end with the definitive closure of older units. Therefore, future electricity prices and investments have to be treated simultaneously, thinking of a transition towards a long-term price equilibrium. This basic idea will be developed in the following sections starting with the static, deterministic long-term equilibrium in section 1. It has been shown that this long-term equilibrium corresponds to the one derived from the concept of peak-load pricing already developed in regulated markets. Sections 2 and 3 deal with stochastic extensions to the peak-load-pricing concept, with the focus being on short-term, i.e. mostly load fluctuations in section 2, and on longer term uncertainties in section 3, looking particularly at varying fuel (or emission) prices. Section 4 then develops a backward induction approach, similar to those used in real option approaches, to derive simultaneously dynamic stochastic price equilibria and the value of different power plants in the long-term perspective. This approach is then applied in section 5.

1. STATIC, DETERMINISTIC LONG-TERM MARKET EQUILIBRIUM - THE CONCEPT OF PEAK LOAD PRICING

Companies will always attempt to shape their decisions with long-term impacts so as to move from the short-term marginal cost curve to the long-term marginal cost curve (see Varian 1992). The latter constitutes an envelope and lower bound of all short-term (i.e. fixed capital stock) cost curves (cf. Figure 9-1), since at different price ratios, the minimum costs achievable at fixed capital stock will at best correspond to the minimum costs achievable when the capital stock is flexible, too. The expression "long-term" is therefore not associated with a precise time span but rather designates a time span sufficiently long so that all restrictions through existing capital stocks have vanished. If usual power plant lifetimes are taken to be 40 years, then the long-term would be at maximum 40 years away. But in virtue of the argument given above, the long-term equilibrium will already influence the investment decisions of today.

Electricity, at least at the wholesale level, is a rather homogenous good. Nevertheless, it has not a uniform price as discussed in chapter 4, notably due to its non-storability and the resulting necessity for a supply and demand balance at each single point in time. As shown in chapter 4, section 1.1, the equilibrium price will at each point in time be equal to the marginal generation costs of the last unit needed to fulfill the demand restriction. This argument can be directly generalized to the long-term equilibrium case with the slight, but important difference, that the last unit needed in the peak load time, has not only to recover its variable generation costs but also its fixed costs, because otherwise it would not be built. The same holds obviously for the other generation technologies.

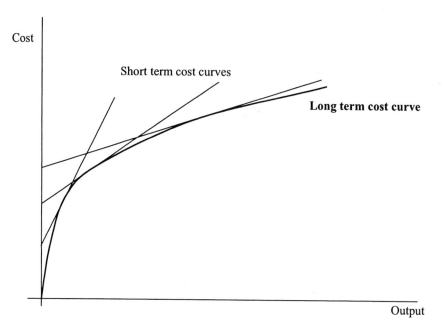

Figure 9-1. Long-term cost functions as lower bound to short-term cost functions

Hence, the problem defined in Eqs. (4-1) to (4-4) for the short-term has to be extended, making capacities not a right-hand side constraint but a decision variable. This requires notably the inclusion of the (annualized) capital costs $C_{Inv,u}$ into the objective formulation and correspondingly a shift of the capacity parameters from the right to the left side of the inequalities:

$$\min_{y_{u,t},K_u} C \qquad (9\text{-}1)$$

$$C = \sum_t \sum_u C_{Op,u,t} + \sum_u C_{Inv,u} = \sum_t \sum_u p_{F,u,t} h_u y_{u,t} + \sum_u a(\rho, L_u) c_{Inv,tot,u} K_{PL,u}$$
$$= \sum_t \sum_u c_{Op,u} y_{u,t} + \sum_u c_{Inv,u} K_{PL,u}$$

$$\qquad (9\text{-}2)$$

$$\sum_u y_{u,t} \geq D_t \qquad (9\text{-}3)$$

$$y_{u,t} - K_{PL,u} \leq 0 \qquad (9\text{-}4)$$

As in the short-term case, demand elasticity is assumed to be zero. It has been shown (cf. e.g. Oren 2000 for a short summary) that the optimal solution to the technology choice problems can be obtained graphically as shown in Figure 9-2. In the lower part, the total (annualized) costs per MW c_u are plotted as a function of total annual operation hours $t_{Op,u}$:

$$t_{Op,u} = \left(\sum_t y_{u,t}\right)\bigg/K_{PL,u} \qquad\qquad (9\text{-}5)$$

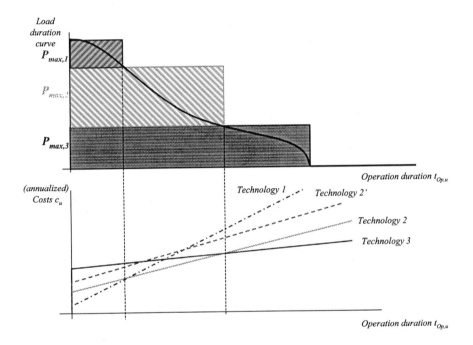

Figure 9-2. Graphical solution to the peak-load pricing problem

From Eq. (9-2), it is then obvious that (at time-invariant fuel prices):

$$c_u = C_u/K_u = p_{F,u}h_u t_{Op,u} + a(\rho, L_u) \qquad\qquad (9\text{-}6)$$

Thus c_u is a linear function in $t_{Op,u}$ with the capital costs $a(\rho, L_u)$ representing the intersection point on the vertical axis and the variable costs $p_{F,u}\,h_u$ corresponding to the slope of the line. In the example depicted in Figure 9-2, clearly technology 1 is the cheapest for operation duration between 0 and t_1, technology 2 is cheapest between t_1 and t_2 and technology 3 is then the most economical for all operation times exceeding t_2. Since in the upper part, the load in all time segments t has been ordered by decreasing values yielding the so-called load duration curve, this may be directly used to determine the optimal installed capacity for each technology. Starting with the highest load duration, the optimal capacity for technology 3 (the baseload plants) is determined as the load level which is exceeded in at least t_2 hours per year. In statistical terms, the $(1 - t_2/8760)$-quantile of the load duration curve is chosen as optimal capacity $K_{PL,2}$. By similar reasoning, the optimal capacity for technology 2 is determined as the difference between the $(1 - t_1/8760)$- and the $(1 - t_2/8760)$-quantile of the load duration curve. And finally the peak load technology 1 has to have an equilibrium capacity equal to the difference between the maximum load x_{max} and the $(1 - t_1/8760)$-quantile.

A technology, which at all possible numbers of operating hours is more expensive than another one, will not be selected as part of the efficient portfolio. This is the case of technology 2' in Figure 9-2: the cost curve is throughout above the cost curve for technology 2, so it is said to be "dominated" by technology 2 and it would be inefficient to use it. Formally, a technology u' is inefficient if

$$\forall t_{Op} \, \exists u \; c_u\!\left(t_{Op}\right) < c_{u'}\!\left(t_{Op}\right) \tag{9-7}$$

i.e., for all operating hours $t_{Op,u}$, there exists another technology u (not necessarily always the same) which is less costly than u. Such technologies can be excluded a priori from the optimization problem at given prices.

As in chapter 4, section 1, prices are mathematically obtained as solutions of the dual problem corresponding to the one formed by Eqs. (9-1) to (9-4). By the same reasoning, as long as capacities are not scarce, at least for one type of plant, prices will not exceed the variable costs of the last unit used for satisfying demand. If the capacity constraint of the last unit gets binding, then (shadow) prices will get up to the level of annualized investment costs for the peaking unit (cf. Eq. 4-8). If several hours with the same peak load level exist, the investment costs would be divided evenly between these hours, but in the case of a single deterministic peak load hour, this single hour would bear theoretically the whole capacity costs. Graphically, the prices correspond to the derivative of the efficient cost frontier $c^*(t_{Op})$ and the price peak at hour 0 (or hour 0 to n) arises from the step in the cost functions at $t = 0$, which corresponds to the capital costs of the technologies.

2. STOCHASTIC FLUCTUATIONS IN THE ELECTRIC LOAD

The analysis presented in the previous section has been extended in a series of papers starting with Brown and Johnson (1969) and including Visscher (1973), Crew and Kleindorfer (1976), Carlton (1977) and Chao (1983) to the case of stochastic demands. Albeit the historical starting point has been utility planning, not all authors distinguish diverse technologies. Therefore the subsequent presentation follows the treatment by Crew and Kleindorfer (1976) and Chao (1983), which explicitly account for diverse technologies.

Under uncertainty, the starting point is in general welfare maximization in a partial equilibrium context instead of cost minimization, since neglecting or negating demand elasticity leads automatically to a problem setting, where it is optimal to provide supply capacities capable of coping with any conceivable load, i.e. corresponding to the maximum load.

Welfare to be maximized includes consumer surplus plus total revenue minus production costs. Some authors such as Crew and Kleindorfer (1976) also include rationing costs to cope with cases where supply curtailment can not be applied to those customers with the lowest willingness to pay for electricity. Many authors including Brown and Johnson (1969), Crew and Kleindorfer (1976) and Carlton (1977) formalize welfare as a function of prices and capacities, looking then for the necessary (and mostly sufficient) conditions for a welfare optimum. However, this conceptualization mixes primal and dual variables from the outset, and the prices obtained are then conditional on the preset capacities, hence not including capacity costs. Therefore, the welfare functional and the maximization problem are formulated in primal variables (production quantities $y_{u,t}(\varepsilon_t)$ and capacities K_u) in the following and prices are derived as shadow variables, as in the cost minimization problems dealt with in section 1 and chapter 4, section 1. Taking expectations over all possible values of the stochastic variables $\varepsilon = \{\varepsilon_t\}$,(thus risk neutrality is assumed), the welfare functional $\overline{W}$ to be maximized is:

$$\overline{W}\big(\mathbf{y}(\varepsilon),\mathbf{y}_{tot}(\varepsilon),\mathbf{K}_{PL}\big) = E_\varepsilon\big[W\big(\mathbf{y}(\varepsilon),\mathbf{y}_{tot}(\varepsilon),\mathbf{K}_{PL}\big)\big] \qquad (9\text{-}8)$$

with the welfare in each scenario ε equal to:

$$W\big(\mathbf{y}(\mathbf{\varepsilon}),\mathbf{y}_{tot}(\mathbf{\varepsilon}),\mathbf{K}_{PL}\big)=\sum_{t}\int_{0}^{y_{tot,t}(\varepsilon_t)}V\big(\widetilde{y}_t,\varepsilon_t\big)d\widetilde{y}_t-\sum_{t}\sum_{u}c_{Op,u}y_{u,t}(\varepsilon_t)-\sum_{u}c_{Inv,u}K_u$$

$$\therefore\qquad y_{u,t}(\varepsilon_t)\le K_u$$

$$y_{tot,t}(\varepsilon_t)\le\sum_{u}y_{u,t}(\varepsilon_t)$$

(9-9)

Thereby, $V\big(\widetilde{y}_t,\varepsilon_t\big)$ is the value (willingness-to-pay) attached by the customers to quantity $\widetilde{y}_t$ of total electricity produced in a scenario with stochastic value ε_t. It thus corresponds to an inverse demand function. $c_{Op,u}=p_{F,u}\,h_u$ are the operating costs per output $y_{t,\varepsilon u}$ and $c_{Inv,u}$ the annualized investment costs. The restrictions can be included in the overall problem by introducing the Lagrangian multiplier functions $\psi_{S,t}(\varepsilon_t)$, interpreted as the shadow prices of supply at time t as a function of scenario ε_t, and $\psi_{K,u,t}(\varepsilon_t)$, which corresponds to the shadow prices of the capacity restriction at time t. Maximizing $\overline{W}\big(\mathbf{y}(\mathbf{\varepsilon}),\mathbf{y}_{tot}(\mathbf{\varepsilon}),\mathbf{K}_{PL}\big)$ is then equivalent to maximizing the Lagrangian $\widetilde{W}\big(\mathbf{y}(\mathbf{\varepsilon}),\mathbf{y}_{tot}(\mathbf{\varepsilon}),\mathbf{K}_{PL},\mathbf{\psi}_S(\mathbf{\varepsilon}),\mathbf{\psi}_K(\mathbf{\varepsilon})\big)$ with:

$$\widetilde{W}\big(\mathbf{y}(\mathbf{\varepsilon}),\mathbf{y}_{tot}(\mathbf{\varepsilon}),\mathbf{K}_{PL},\mathbf{\psi}_S(\mathbf{\varepsilon}),\mathbf{\psi}_K(\mathbf{\varepsilon})\big)=\overline{W}\big(\mathbf{y}(\mathbf{\varepsilon}),\mathbf{y}_{tot}(\mathbf{\varepsilon}),\mathbf{K}_{PL}\big)$$

$$-\sum_{t}\int_{-\infty}^{-\infty}\psi_{S,t}(\varepsilon_t)\left(y_{tot,t}(\varepsilon_t)-\sum_{u}y_{u,t}(\varepsilon_t)\right)f(\varepsilon_t)d\varepsilon_t$$

$$-\sum_{t}\int_{-\infty}^{-\infty}\psi_{K,u,t}(\varepsilon_t)\big(y_{u,t}(\varepsilon_t)-K_{PL,u}\big)f(\varepsilon_t)d\varepsilon_t$$

(9-10)

Thereby, the Kuhn-Tucker conditions:

$$\psi_{S,t}(\varepsilon_t)\ge0\;\wedge\;\sum_{u}y_{u,t}(\varepsilon_t)-y_{tot,t}(\varepsilon_t)\le0\;\wedge\;\psi_{S,t}(\varepsilon_t)\left(\sum_{u}y_{u,t}(\varepsilon_t)-y_{tot,t}(\varepsilon_t)\right)=0$$

(9-11)

$$\psi_{K,u,t}(\varepsilon_t)\ge0\;\wedge\;y_{u,t}(\varepsilon_t)-K_{PL,u}\le0\;\wedge\;\psi_{K,u,t}(\varepsilon_t)\big(y_{u,t}(\varepsilon_t)-K_{PL,u}\big)=0 \quad(9\text{-}12)$$

have to be fulfilled.

As necessary conditions for an optimum, we get under mild regularity conditions, applying the principles of variational calculus:

$$\frac{\partial \tilde{W}}{\partial y_{u,t}(\varepsilon_t)} = \psi_{S,t}(\varepsilon_t) - c_{Op,u} \mathbf{1}_{y_{u,t}(\varepsilon_t)>0} - \psi_{K,u,t}(\varepsilon_t) \overset{!}{=} 0 \tag{9-13}$$

$$\frac{\partial \tilde{W}}{\partial y_{tot,t}(\varepsilon_t)} = \frac{\partial}{\partial y_{tot,t}(\varepsilon_t)} \int_0^{y_{tot,t}(\varepsilon_t)} V(\tilde{y}_t, \varepsilon_t) d\tilde{y}_t - \psi_{S,t}(\varepsilon_t) = V(y_{tot,t}(\varepsilon_t), \varepsilon_t) - \psi_{S,t}(\varepsilon_t) \overset{!}{=} 0 \tag{9-14}$$

$$\frac{\partial \tilde{W}}{\partial K_{PL,u}} = -c_{Inv,u} + \sum_t \int_{-\infty}^{-\infty} \psi_{K,u,t}(\varepsilon_t) f(\varepsilon_t) d\varepsilon_t \overset{!}{=} 0 \tag{9-15}$$

$$\frac{\partial \tilde{W}}{\partial \psi_{S,t}(\varepsilon_t)} = \sum_u y_{u,t}(\varepsilon_t) - y_{tot,t}(\varepsilon_t) \overset{!}{\leq} 0 \tag{9-16}$$

$$\frac{\partial \tilde{W}}{\partial \psi_{K,u,t}(\varepsilon_t)} = y_{u,t}(\varepsilon_t) - K_{PL,u} \overset{!}{\leq} 0 \tag{9-17}$$

From Eqs. (9-19) and (9-20) and the Kuhn-Tucker conditions (9-17), it follows, given the non-negativity of the shadow price $\psi_{K,u,t}$ that:

$$V\left(\sum_u y_{u,t}(\varepsilon_t), \varepsilon_t\right) \geq c_{Op,u} \mathbf{1}_{y_{u,t}(\varepsilon_t)>0} \tag{9-18}$$

meaning that the willingness-to-pay of the customers must at least equal the marginal costs of a unit in those periods when the unit is running. Consequently, the units will again be dispatched according to the merit order by increasing operation costs, yet not until a prespecified demand is fulfilled, but rather until the marginal costs of the last running unit equal the marginal willingness-to-pay at an endogenous output level (cf. Figure 9-3).

$\psi_{s,t}$ is independent of the unit considered and may be represented graphically as an intersection between supply and demand curves. Thereby the supply curve is formed by the ordered and cumulated individual capacities, given that the inverse demand function is expressed as a function of cumulative production.

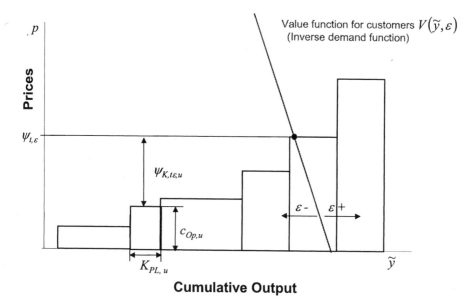

Cumulative Output

Figure 9-3. Shadow prices in welfare maximization under uncertainty

From Eq. (9-19), the price at given ε_t may also be written:

$$\psi_{S,t}(\varepsilon_t) = \begin{cases} c_{Op,u} & \text{if } \hat{K}_{PL,u-1} < y_{tot,t}(\varepsilon_t) < \hat{K}_{PL,u} \\ c_{Op,u} + \psi_{K,u,t}(\varepsilon_t) & \text{if } \hat{K}_{PL,u} = y_{tot,t}(\varepsilon_t) \\ c_{Op,\hat{u}} + \psi_{K,\hat{u},t}(\varepsilon_t) & \text{if } \hat{K}_{PL,\hat{u}} = y_{tot,t}(\varepsilon_t) \end{cases} \qquad (9\text{-}19)$$

Thereby, it is (without limitation of the generality) assumed that the units u are ordered by increasing operational cost from 1 to $\hat{u}$ and the cumulative capacity $\hat{K}_{PL,u}$ up to unit u is defined as:

$$\hat{K}_{PL,u} = \sum_{u'=1}^{u} K_{PL,u'} \qquad (9\text{-}20)$$

If the inverse D_t of the willingness-to-pay function V with respect to quantity y_{tot} is additive:

$$D_t(p_t,\varepsilon_t) = V_t^{-1}(p_t,\varepsilon_t) = \tilde{D}_t(p_t) + \varepsilon_t \qquad (9\text{-}21)$$

the case differentiation in Eq. (9-25) can be expressed directly in terms of the distribution of ε_t using Eq. (9-20).

$$\psi_{S,t}(\varepsilon_t) = \begin{cases} c_{Op,u} & \text{if } \hat{K}_{PL,u-1} - \tilde{D}_t(c_{Op,u}) < \varepsilon_t < \hat{K}_{PL,u} - \tilde{D}_t(c_{Op,u}) \\ c_{Op,u} + \psi_{K,u,t}(\varepsilon_t) & \text{if } \hat{K}_{PL,u} - \tilde{D}_t(c_{Op,u}) \le \varepsilon_t \le \hat{K}_{PL,u} - \tilde{D}_t(c_{Op,u+1}) \\ c_{Op,\hat{u}} + \psi_{K,\hat{u},t}(\varepsilon_t) & \text{if } \hat{K}_{PL,\hat{u}} - \tilde{D}_t(c_{Op,\hat{u}}) \le \varepsilon_t \end{cases}$$

$$(9\text{-}22)$$

The ex-ante average price, which should be charged to customers, is then obtained through integrating over the distribution of ε_t:

$$\begin{aligned} \overline{\psi}_{S,t} = &\sum_{u=1}^{\hat{u}} \left(F_{\varepsilon_t}\left(\hat{K}_{PL,u} - \tilde{D}_t(c_{Op,u+1}) \right) - F_{\varepsilon_t}\left(\hat{K}_{PL,u-1} - \tilde{D}_t(c_{Op,u}) \right) \right) c_{Op,u} \\ &+ \sum_{u=1}^{\hat{u}-1} \int_{\hat{K}_{PL,u} - \tilde{D}_t(c_{Op,u})}^{\hat{K}_{PL,u} - \tilde{D}_t(c_{Op,u+1})} \psi_{K,u,t}(\varepsilon_t) f(\varepsilon_t) d\varepsilon_t \\ &+ \int_{\hat{K}_{PL,\hat{u}} - \tilde{D}_t(c_{Op,\hat{u}})}^{+\infty} \psi_{K,\hat{u},t}(\varepsilon_t) f(\varepsilon_t) d\varepsilon_t \end{aligned}$$

$$(9\text{-}23)$$

Thus, the optimal ex-ante price is composed of three components:
1. a weighted average of the variable costs of the different plants, each plant being weighted with the probability that it will be the last plant needed for the supply-demand balance in the time segment considered
2. the probability weighted shadow prices for the different cumulative capacities which are paid before the next unit is put online
3. the probability weighted shadow prices of the total installed capacity $K_{PL,\hat{u}}$, i.e. the capacity rents obtained on total installed capacity

The deterministic case with inflexible demand is contained as special case in Eq. (9-29). In that case, the variable costs of one single plant are retained with probability 1 in the first term, the probability for all other terms being zero. The second term vanishes due to demand inflexibility and the third term is zero in all periods except the peaking period, where it is set so that the investment costs of the peaking unit are recovered. In the stochastic case, the investment costs for the peaking unit are recovered through the summation over all time steps of the probability weighted capacity rents generated during these time steps. From Eq. (9-21) one gets, taking into consideration Eq. (9-25):

$$-c_{Inv,\hat{u}} + \sum_t \int\limits_{\hat{K}_{PL,\hat{u}} - \tilde{D}_t(c_{Op,\hat{u}})}^{+\infty} \psi_{K,\hat{u},t}(\varepsilon_t) f(\varepsilon_t) d\varepsilon_t \overset{!}{=} 0 \qquad (9\text{-}24)$$

In view of the third case of Eq. (9-28) (in the others, the peaking unit is not producing), we obtain:

$$c_{Inv,\hat{u}} = \sum_t \int\limits_{\hat{K}_{PL,\hat{u}} - \tilde{D}_t(c_{Op,\hat{u}})}^{+\infty} \left(\psi_{S,t}(\varepsilon_t) - c_{Op,\hat{u}} \right) f(\varepsilon_t) d\varepsilon_t \qquad (9\text{-}25)$$

Using the equality (9-20) of shadow price $\psi_{s,t}$ and willingness-to-pay V and the additive decomposition (9-27) of the inverse willingness-to-pay function $V^{-1} = D$, this yields:

$$c_{Inv,\hat{u}} = \sum_t \int\limits_{\hat{K}_{PL,\hat{u}} - \tilde{D}_t(c_{Op,\hat{u}})}^{+\infty} \left(\tilde{D}_t^{-1} \left(\hat{K}_{PL,\hat{u}} - \varepsilon_t \right) - c_{Op,\hat{u}} \right) f(\varepsilon_t) d\varepsilon_t \qquad (9\text{-}26)$$

At given deterministic demand functions $\tilde{D}_t$, stochastic distributions $f(\varepsilon_t)$ and costs $c_{Inv,\hat{u}}$ and $c_{Op,\hat{u}}$, this provides a single equation for determining the total installed capacity $\hat{K}_{PL,\hat{u}}$ given the cost characteristics of the peaking unit. In general, this equation may only be solved numerically for $\hat{K}_{PL,\hat{u}}$. The right-hand side of Eq. (9-32) is monotonously decreasing in $\hat{K}_{PL,\hat{u}}$, since the integrand is decreasing and the lower integration bound is increasing in $\hat{K}_{PL,\hat{u}}$. Thus, Eq. (9-32) has at maximum one solution for $\hat{K}_{PL,\hat{u}}$. The case of no solution at all will only appear if even at zero installed capacity ($\hat{K}_{PL,\hat{u}} = 0$), the willingness to pay is so low that the sum of running and investment costs is not covered and consequently no electricity at all is produced. This seems hardly relevant in practice.

A similar equation as (9-32) can also be derived for the cumulative capacities $\hat{K}_{PL,u}$ for the other units u, taking into consideration all three cases of Eq. (9-28):

$$c_{Inv,u} = \sum_t \left\{ \sum_{u'=u}^{\hat{u}-1} \left(\int_{\hat{K}_{PL,u'} - \tilde{D}_t(c_{Op,u'})}^{\hat{K}_{PL,u'} - \tilde{D}_t(c_{Op,u'+1})} \left(\tilde{D}_t^{-1}(\hat{K}_{PL,u'} - \varepsilon_t) - c_{Op,u} \right) f(\varepsilon_t) d\varepsilon_t \right. \right.$$

$$\left. + \int_{\hat{K}_{PL,u'} - \tilde{D}_t(c_{Op,u'+1})}^{\hat{K}_{PL,u'+1} - \tilde{D}_t(c_{Op,u'+1})} \left(c_{Op,u'+1} - c_{Op,u} \right) f(\varepsilon_t) d\varepsilon_t \right) \qquad (9\text{-}27)$$

$$\left. + \int_{\hat{K}_{PL,\hat{u}} - \tilde{D}_t(c_{Op,\hat{u}})}^{+\infty} \left(\tilde{D}_t^{-1}(\hat{K}_{PL,\hat{u}} - \varepsilon_t) - c_{Op,u} \right) f(\varepsilon_t) d\varepsilon_t \right\}$$

Each term in the summation of Eq. (9-33) corresponds to the intersection of the willingness-to-pay function V with a different segment of the cumulative supply function (cf. Figure 9-3). The integration bounds describe the conditions to be fulfilled by ε_t to match this case (corresponding to the bounds set in Eq. 9-28), whereas the integrand then gives the value of the shadow price $\psi_{K,u,t}$ in this particular case, corresponding to the marginal supply price minus the operating cost of the unit u considered. Equation (9-33) can also be written recursively, yielding:

$$c_{Inv,u} = \sum_t \left\{ \int_{\hat{K}_{PL,u} - \tilde{D}_t(c_{Op,u})}^{\hat{K}_{PL,u} - \tilde{D}_t(c_{Op,u+1})} \left(\tilde{D}_t^{-1}(\hat{K}_{PL,u} - \varepsilon_t) - c_{Op,u} \right) f(\varepsilon_t) d\varepsilon_t \right.$$

$$\left. + \int_{\hat{K}_{PL,u} - \tilde{D}_t(c_{Op,u+1})}^{+\infty} \left(c_{Op,u} - c_{Op,u+1} \right) f(\varepsilon_t) d\varepsilon_t \right\} + c_{Inv,u+1} \qquad (9\text{-}28)$$

Thus the differences in investment costs between technology u and $u+1$ have to be refinanced through the corresponding savings in operation costs, when both u and $u+1$ are running (second term) plus the capacity rent obtained in those cases when the capacities of u are fully used but the willingness-to-pay does not permit to operate also $u+1$ (first term). From Eq. (9-34), the installed cumulative capacities up to unit u can be determined recursively, starting with the peaking unit $\hat{u}$ and working backwards to the baseload unit 1. For each u, Eq. (9-34) has again to be solved numerically. The capacities for the technologies themselves can then obtained by simple subtraction:

$$K_{PL,u} = \hat{K}_{PL,u} - \hat{K}_{PL,u-1} \qquad (9\text{-}29)$$

The deterministic, price-inelastic case again is contained as a special case in Eq. (9-34). The first integral vanishes in that case since lower and upper integration boundary coincide and for the evaluation of the second the Dirac impulse as a limiting distribution for the density function $f(\varepsilon_t)$ is taken (cf. chapter 4). Then the second integral equals 0 if the cumulative capacity $\hat{K}_{PL,u}$ exceeds current demand $\tilde{D}_t$ and corresponds to the difference in operation costs otherwise. Thus adding up over all time steps yields:

$$c_{Inv,u} = t_{Op,u+1}\left(c_{Op,u} - c_{Op,u+1}\right) + c_{Inv,u+1} \qquad (9\text{-}30)$$

with $t_{Op,u+1}$ the operation hours left for the more expensive unit $u+1$, once capacity $\hat{K}_{PL,u}$ of cheaper units has been deployed. As can easily be verified, Eq. (9-36) precisely describes the location of the intersection point between the cost curves in the middle part of Figure 9-2[58].

3.   STOCHASTIC FLUCTUATIONS IN THE PRIMARY ENERGY PRICES

Besides uncertainties in the load, the long-term market equilibrium is also affected by fuel price uncertainty[59]. Fuel prices affect the operation costs of the plants, and thus both prices and optimal capacities in a long-term equilibrium depend on observed or expected fuel prices (cf. Eqs. 9-29, 9-34, 9-36). Fuel price uncertainty arises especially in the longer run, thus in the following no uncertainties within the planning period are considered. Rather, fuel prices - and consequently operating costs – are considered as equal in all time segments within the planning year considered.

Two basic cases for the information-decision structure can then be envisaged: in the first case, prices for fuels are revealed before the decisions on capacities are made. In this case, for each fuel price scenario, different

---

[58] Another particular case, which is also covered by the general stochastic model, is peak-shifting. Peak-shifting arises in the model especially when it is specified as a deterministic model but with elastic demand. Contrarily to section 1, allocating the whole capacity price to the peak segment would then depress the demand in this segment so far that it would fall beyond the demand in other time segments. According to equation (9-23), the price in each time segment will rather take into account the capacity rent obtained by each unit even outside the peak period.

[59] Furthermore, uncertainties on plant availabilities may also affect the long-term market equilibrium (cf. Figure 3-1). But these shall not be considered in detail here. Technological uncertainties have already been discussed in section 8.

optimal capacities are determined with the same methods as used in section 1 or section 2. Less obvious is the situation in the second case: investment decisions are made before fuel prices are revealed. This corresponds to the actual situation given the considerable lead times between decisions to invest and first power production in the new plant.

In this case the basic analysis carried out in the previous section is still valid, especially the welfare functional and the first order-conditions derived in Eqs. (9-14) to (9-23) are unchanged, except that now the operation costs $c_{Op,u}$ have to be written as a function of the stochastic variables ε_f, corresponding to the stochastic part of the fuel prices p_f. For the market price, the following conditions analogous to Eq. (9-28) are then obtained:

$$\psi_{S,t}(\varepsilon_t) = \begin{cases} c_{Op,u}(\varepsilon_{f(u)}) & \text{if } \hat{K}_{PL,u-1|\varepsilon} < \tilde{D}_t(c_{Op,u}(\varepsilon_{f(u)})) < \hat{K}_{PL,u} \\ c_{Op,u}(\varepsilon_{f(u)}) + \psi_{K,u,t}(\varepsilon) & \text{if } c_{Op,u}(\varepsilon_{f(u)}) \leq \tilde{D}_t^{-1}(\hat{K}_{PL,u}) \leq c_{Op,u+1|\varepsilon}(\varepsilon_{f(u)}) \\ c_{Op,\hat{u}}(\varepsilon_{f(u)}) + \psi_{K,\hat{u},t}(\varepsilon) & \text{if } \hat{K}_{PL,\hat{u}} = \tilde{D}_t(c_{Op,\hat{u}|\varepsilon}(\varepsilon)) \end{cases}$$

$$(9\text{-}31)$$

Here, an unambiguous ordering of the different units according to their operational cost is not always possible. Depending on the scenario ε, varying units are possibly most expensive (or cheapest). This is expressed by making the indices of the preceding unit u-1, the succeeding unit u+1 and the peaking unit $\hat{u}$ conditional on the scenario ε.

Correspondingly, the ex-ante average price that should be charged to customers can only be obtained by integrating over the stochastic vector ε and by inverting the sequence of integration and summation compared to Eq. (9-29):

$$\overline{\psi}_{S,t} = \sum_u \left(F_\varepsilon\left(\hat{K}_{PL,u} - \tilde{D}_t\left(c_{Op,u+1|\varepsilon}(\varepsilon_{f(u+1)})\right)\right) - F_\varepsilon\left(\hat{K}_{PL,u-1|\varepsilon} - \tilde{D}_t\left(c_{Op,u}(\varepsilon_{f(u)})\right)\right) \right) c_{Op,u}(\varepsilon_{f(u)})$$

$$+ \int\limits_{\substack{\varepsilon \\ \psi_{K,u,t}(\varepsilon) \leq c_{Op,u+1|\varepsilon} - c_{Op,u}}} \sum_u \psi_{K,u,t}(\varepsilon) f(\varepsilon) d\varepsilon$$

$$(9\text{-}32)$$

Since the peaking unit is not known with certitude, the second and third summation terms in Eq. (9-29) have also been condensed into one term.

This term also corresponds to the capacity rents used to recover the investment costs. In analogy to Eq. (9-33), one gets:

$$c_{Inv,u} = \sum_t \int\limits_{\substack{\varepsilon \\ \psi_{K,u,t}(\varepsilon) \le c_{Op,u+1|\varepsilon} - c_{Op,u}}} \psi_{K,u,t}(\varepsilon) f(\varepsilon) d\varepsilon \tag{9-33}$$

This multiple integral is obviously not easily solved in the general case. Only in the case when the fuel price changes do not lead to changes in the merit order, a similar recursive procedure to the one described in section 2 can be used. If the (small) consequences of changes in operation costs on total demand are disregarded, the optimal investment capacities may even be determined by using average fuel prices:

$$
\begin{aligned}
c_{Inv,u} = \sum_t & \left\{ \sum_{u'=u}^{\hat{u}-1} \left(\int_{-\infty}^{+\infty} 1_{\tilde{D}_t(c_{Op,u'+1}(\varepsilon_{f(u'+1)})) \le \hat{K}_{PL,u'} \le \tilde{D}_t(c_{Op,u'}(\varepsilon_{f(u')}))} \left(\tilde{D}_t^{-1}(\hat{K}_{PL,u'}) - c_{Op,u} \right) f(\varepsilon_{f(u')}) d\varepsilon_t \right. \right. \\
& \left. + \int_{-\infty}^{+\infty} \int_{-\infty}^{+\infty} 1_{\hat{K}_{PL,u'-1} < \tilde{D}_t(c_{Op,u'}(\varepsilon_{f(u')})) \le \hat{K}_{PL,u'}} \left(c_{Op,u'} - c_{Op,u} \right) f(\varepsilon_{f(u')}, \varepsilon_{f(u)}) d\varepsilon_t \right) \\
& \left. + \int_{-\infty}^{+\infty} 1_{\hat{K}_{PL,\hat{u}} = \tilde{D}_t(c_{Op,u'}(\varepsilon_{f(u')}))} \left(\tilde{D}_t^{-1}(\hat{K}_{PL,\hat{u}}) - c_{Op,u} \right) f(\varepsilon_t) d\varepsilon_t \right\} \\
\approx \sum_t & \left\{ \sum_{u'=u}^{\hat{u}-1} \left(1_{\hat{K}_{PL,u'-1} < \tilde{D}_t(c_{Op,u'}) < \hat{K}_{PL,u'}} \left(\bar{c}_{Op,u'} - \bar{c}_{Op,u} \right) \right) + 1_{\hat{K}_{PL,\hat{u}} \le \tilde{D}_t(c_{Op,\hat{u}})} \left(\tilde{D}_t^{-1}(\hat{K}_{PL,\hat{u}}) - \bar{c}_{Op,u} \right) \right\}
\end{aligned}
$$

$$\tag{9-34}$$

Thus, fuel price uncertainty can be neglected in the case of no shifts in merit order and inflexible demands and instead the average expected values of operation costs may be used. Since the probability for reversals in the merit order is usually very limited for short and medium term time lags between investment decision and operation, a use of Eq. (9-34) is justified for computing static "long-term" market equilibria under fuel price uncertainty. But these equilibria are long term only in the sense that they do not treat capital investments as fix but as variable. They do not reflect the longer run risk of price changes, which may decrease the profits for any envisaged investment over its lifetime. The dynamic model of the next section will deal with these effects.

4. DYNAMIC STOCHASTIC LONG-TERM PRICE EQUILIBRIA - A BACKWARD INDUCTION APPROACH

The major market risk for any power plant investment in the longer run is that fuel prices (and/or technology) develop in a way that a once-built power plant is not competitive any more. Thereby two cases have to be distinguished: one possibility is that the technology is no longer part of the efficiency frontier at all. Another is that the range of efficient operation hours (and consequently the optimally installed capacity) of the technology decreases. In both cases, the capacities already installed can still be operated, but they have to accept a reduced operation margin (cf. Figure 9-4).

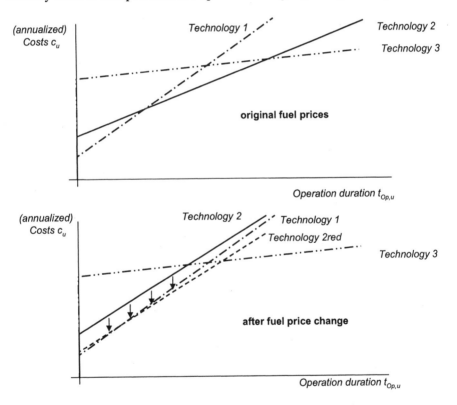

Figure 9-4. Impact of primary energy price changes on competitiveness of generation technologies and resulting return on investment

Yet, if on the contrary the range of efficient operation hours increases, this does not induce any extra profit for units already installed. It creates rather an incentive to install more units of the same type. Hence, if such a fuel price uncertainty exists, the owner of the power plant will require a higher return on investment in the first period as compensation for the risks incurred in later periods.

But in order to derive valuable results for the single investor, it is obviously necessary to treat the whole industry equilibrium. Dixit and Pindyck (1994) develop general methods to deal with the investment-price-equilibrium under uncertainty at the industry level and they also treat several application examples taken from the electricity industry, but they do not analyze the practically relevant setting which includes simultaneously fuel switching, fuel price uncertainty and endogenous output prices.

Obviously, in this setting an analytical solution to the dynamic price-investment equilibrium is hardly possible, given that fuel switching turned out to complicate already the static case of section 3. Therefore a numerical approach is developed in the following. Firstly, the general problem formulation is given in section 4.1, then an illustrative example is discussed in section 4.2 and a general solution methodology developed in section 0. Finally a larger application is described in section 5.

4.1   Basic problem

Dixit and Pindyck (1994) derive the industry equilibrium from an analysis of the optimal decisions of individual firms. But they then show that the industry equilibrium is also welfare optimal. Therefore, an equivalent approach is to formulate the welfare optimization problem for investment under uncertainty. As in most of the analysis of Dixit and Pindyck, issues of market power are neglected in the following. As starting point the stochastic but static problem formulations discussed in the previous sections are taken. To simplify somewhat, demand uncertainty and demand price elasticity are neglected as in section 1 and 3, so that the problem of welfare optimization is equivalent to a problem of cost minimization.

In the dynamic setting, now a double time index has to be considered: the first one, labeled T, is running over the planning periods and the second one t is running over the hours within the planning period. Furthermore, a scenario index s has to be added to distinguish the different possible states of the world (reflecting notably different fuel prices).

Formulating the problem in discrete time in view of a later numerical treatment, the cost function to be minimized is (analogously to Eqs. 9-14 and 9-15):

$$C_{T,s}\left(\mathbf{K}_{\mathrm{PL},T,s}, \mathbf{y}, \mathbf{y}_{tot}; \widetilde{\mathbf{K}}_{\mathrm{PL},T-1,s_0}\right) = b(\rho,1,d_T)\sum_{t}\sum_{u}c_{Op,u,1,s}\,y_{u,T,s,t} + \sum_{u}c_{Inv,u}K_{T,s,u} +$$

$$b(\rho,d_T,d_T)\sum_{s'}\mathrm{Pr}_{s\to s'}\,C_{T+1,s}{}^{*}\left(\widetilde{\mathbf{K}}_{\mathrm{PL},T,s}\right)$$

$$\therefore \qquad y_{u,T,s,t} \le \widetilde{K}_{\mathrm{PL},T,s,u}$$

$$y_{tot,T,s,t} \le \sum_{u}y_{u,T,s,t}$$

$$\widetilde{K}_{\mathrm{PL},T,s,u} \le \widetilde{K}_{\mathrm{PL},T-1,s_0,u} + K_{T,s,u} - \zeta_{T-1\to T}\left(\widetilde{K}_{\mathrm{PL},T-1,s_0,u}\right)$$

$$(9\text{-}35)$$

Hence the sum of the costs for operating the system in the current period including necessary investments plus the probability weighted sums of minimal costs for the subsequent periods are minimized under the capacity and the load constraint for each time segment. The third side constraint describes the evolution of total capacities $\widetilde{K}_{\mathrm{PL},T,s,u}$ per plant type u as a result of investments $K_{T,s,u}$ and scrapping $\zeta_{T-1\to T}\left(\widetilde{K}_{\mathrm{PL},T-1,s_0,u}\right)$. For scrapping, different rules can be envisaged, such as the exponential decay and the finite lifetime ("sudden death") rules considered by Dixit and Pindyck (1994), or the logistic vintage specific decay rule investigated by Schuler (2000).

In view of a flexible implementation, the duration d_T of the planning period T can be one year or more, assuming that operation during these years is similar. The present value factor $b(\rho,1,d_T)$ is therefore used in Eq. (9-35) as a multiplicative factor on the operation costs. The general definition used here is:

$$b(\rho,d_1,d_2) = \sum_{\tau=d_1}^{d_2}\frac{1}{(1+\rho)^{\tau}} \qquad\qquad (9\text{-}36)$$

This allows to use directly the factor $b(\rho,d_T,d_T)$ for discounting from the period $T+1$ to T. If the scenarios s, s' etc. span up the total uncertainties in the decision problem and if the outcomes in the different scenarios can be hedged perfectly on the financial markets, then the interest rate ρ should be set equal to the risk free interest rate. In all other cases one might chose a higher ρ in order to represent unaccounted risk factors and/or risk aversion of the decision makers (cf. Dixit and Pindyck 1994). A lower ρ (going possibly as low as 0) may be justified in cases where natural resource depletion and intergenerational welfare considerations are included (cf. Pearce 1998).

To complete the recursive problem formulation, the cost function (9-35) has to be complemented by the formulation of the optimal cost function C_T^*:

$$C_{T,s}^* \left(\widetilde{\mathbf{K}}_{\mathrm{PL},T-1,s_0} \right) = \min_{\mathbf{K}_{\mathrm{PL},T,s}, \mathbf{y}, \mathbf{y}_{tot}} C_{T,s} \left(\mathbf{K}_{\mathrm{PL},T,s}, \mathbf{y}, \mathbf{y}_{tot}; \widetilde{\mathbf{K}}_{\mathrm{PL},T-1,s_0} \right) \qquad (9\text{-}37)$$

Equations (9-37) and (9-35) taken together provide a Bellman equation which can be used for solving the optimal investment problem through backward induction.

In order to obtain a unique solution, terminal conditions have to be added to the problem. Furthermore the state variables have to be looked at in more detail, to be able to make a meaningful numerical discretization. The issues arising here are discussed first on the two-stage example in the next section, before a general solution approach is proposed.

4.2   Two-stage example

For illustrative purposes a two-stage problem with three technologies is considered:

- A gas turbine with 25 % annual efficiency and 230 €/MW investment costs
- A gas-fired combined cycle plant with 57 % efficiency and 430 €/MW investment costs
- A coal-fired plant with 43 % efficiency and 1000 €/MW investment costs

    Since gas prices are much more volatile we limit ourselves to the consideration of gas price risk. Therefore a binomial tree as shown in Figure 9-5 is constructed, considering one single scenario for the first stage and two for the second. The initial gas price is taken to be 12 €/MWh and it is expected to rise either to 18 €/MWh or to decrease to 6 €/kW. The coal price is assumed to be 6 €/MWh throughout. Figure 9-5 shows the cost curves and the installed capacities obtained under these assumptions for the static long-term equilibria in the different nodes. Clearly, the coal technology is dominating in the base scenario and even more in the high gas price scenario, whereas it is inefficient in the case of low gas prices. Thereby a discount rate of 4 % is used and the lifetime of all plants is assumed to be 40 years, yielding an annuity factor of 0.0505. The peak demand is assumed to be 80,000 MW and the minimum (baseload) demand is set to 20,000 MW. Within this range, the load duration curve is assumed to be linear, so that differences in operation time directly translate into differences in load resp. capacity.

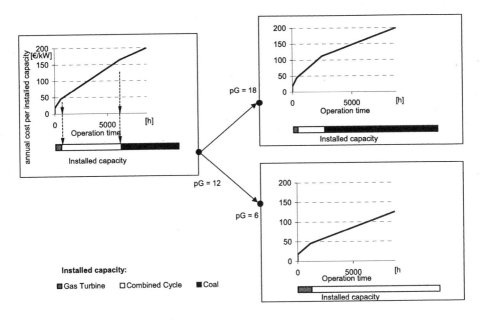

Figure 9-5. Scenario tree for two-stage stochastic investment problem and static equilibria in the nodes

The stochastic cost minimization problem for the first stage can then readily be written using Eqs. (9-37) and (9-35):

$$C_1^*\left(\widetilde{\mathbf{K}}_{\mathbf{PL},0}\right)= \min_{K_{1,u},y_{u,1,t},y_{tot,1,t}} b(\rho,1,d_1)\sum_t\sum_u c_{Op,u,1}y_{u,1,t} +\sum_u c_{Inv,u}K_{1,u} +$$
$$b(\rho,d_1,d_1)\sum_s \Pr_s C_{2,s}^*\left(\widetilde{\mathbf{K}}_{\mathbf{PL},1}\right)$$

$$\therefore \quad y_{u,1,t} \le \widetilde{K}_{\mathbf{PL},1,u}$$
$$y_{tot,1,t} \le \sum_u y_{u,1,t}$$
$$\widetilde{K}_{\mathbf{PL},1,u} \le \widetilde{K}_{\mathbf{PL},0,u} + K_{1,u} -\zeta_{0\to1}\left(\widetilde{K}_{\mathbf{PL},0,u}\right)$$

$$(9\text{-}38)$$

The results will obviously depend on the initial endowments $\widetilde{K}_{\mathbf{PL},0,u}$, on the assumed decay function $\zeta_{0\to1}\left(\widetilde{K}_{\mathbf{PL},0,u}\right)$ and on the cost function $C_{2,s}^*\left(\widetilde{\mathbf{K}}_{\mathbf{PL},1}\right)$ for the second time step. For this second and final time step, the following basic alternatives for the formulation of the cost function may be envisaged[60]:

- Neglect of costs in subsequent periods,
- Static, deterministic formulation of the cost minimization problem in the final period,
- Fixed point solution to an infinite horizon stationary stochastic optimization problem.

The first solution is clearly not very satisfactory, since it implies that investments in the second time step have to earn all their pay back within the second period and this biases clearly the second stage decisions towards those with low capital costs. The second alternative will also tend to bias investment decisions away from flexible, low risk alternatives towards those privileged by conventional net present value calculations. But it is a clear improvement over the first approach. The third solution (cf. Dixit and Pindyck 1994, p. 253f) is theoretically most convincing. But it requires stationary price processes - which is according to the models derived in chapter 4 mostly not the case for primary energy prices. And even if prices are stationary, the solution to the fixed point problem is in general tedious to find, given that a shortcut like the one used by Dixit and Pindyck is not available in most cases. Therefore, the second alternative is pursued in the following.

This leads to the following optimal cost function for the second stage:

$$C_{2,s}^*\left(\tilde{K}_{PL,1}\right) = \min_{K_{2,s,u},y_{u,2,s,t},y_{tot,2,s,t}} b(\rho,1,d_2)\sum_t\sum_u c_{Op,u,2,s}\,y_{u,2,s,t} + \sum_u\left(1-\frac{b(\rho,d_2,L_u)}{b(\rho,1,L_u)}\right)c_{Inv,u}K_{2,s,u}$$

$$\therefore \qquad y_{u,2,s,t} \leq \tilde{K}_{PL,2,s,u}$$

$$y_{tot,2,s,t} \leq \sum_u y_{u,2,s,t}$$

$$\tilde{K}_{PL,2,s,u} \leq \tilde{K}_{PL,1,u} + K_{2,s,u} - \zeta_{1\to2}\left(\tilde{K}_{PL,1,u}\right)$$

(9-39)

Thereby the investment costs for the capacities installed in the second period are reduced by the share $b(\rho,d_2,L_u)/\,b(\rho,1,L_u)$[61]. This share corresponds to the capital pay back to be obtained in the remaining lifetime

---

[60] cf. e. g. the discussions in Kühner (1999) and Rüffler (2000) on terminal constraints in linear programming formulations of optimal investment problems.

[61] An equivalent approach is used by Rüffler (2001) in his linear programming IntE³gral model.

after the end of T_2 under the assumption that the yearly capital pay back is constant, i.e. the technology does not get inefficient after the end of the planning horizon.

Besides this assumption, also the initial capital stock $\tilde{K}_{PL,0,u}$ and the decay functions $\zeta_{0\rightarrow1}\left(\tilde{K}_{PL,0,u}\right)$ and $\zeta_{1\rightarrow2}\left(\tilde{K}_{PL,1,u}\right)$ will influence the results obtained. For the finite lifetime assumption, which is a popular choice for decay functions in energy system models, an additional complication arises given that the function value is not only dependent on the total capacity installed at time 0 resp. 1, but also on when this capacity was installed. Thus, the status variables to be tracked in the backward induction approach are not solely the total installed capacities by technology, but rather the installed capacities by technology and vintage. The same holds for the vintage specific logistic decay model used by Schuler (2000) and other general decay functions used in general energy system model formulations (cf. e.g. Kühner 1996, Rüffler 2001). In the exponential decay model, on the contrary the capacity available at $T+1$ only depends on the capacity available at T and the decay parameter $\lambda=1/L_u$. Yet this assumption seems not very realistic in a world with evolving technologies and subsequent scrapping of the technologies according to their vintage.

Yet a lower bound for the costs can certainly be derived by assuming that $\tilde{K}_{PL,0,u}=0$ so that all capacity is chosen with at least the information of T_1 available. Implicitly the use of information at T_1 corresponds to the assumption that the period T_1 has a duration as long as the lifetime of the longest-living installation, so that the capital stock has been entirely renewed during T_1. From the end of T_1 to the end of T_2, capacities installed in T_1 will then continuously be scrapped. The average decay factor for T_2 is then the average share of capacity scrapped over all the years within T_2. This setting is labeled "with anticipation and full adjustment", since all capacity is freely chosen at T_1 with anticipation of the price distribution at T_2.

An upper bound for the costs is obtained by assuming that the initial capacity $\tilde{K}_{PL,0,u}$ is either known from historical data or chosen with myopic foresight under some operation costs $c_{Op,0,u}$. This setting is labeled "without anticipation and partial adjustment" since the initial capacity choice does not account for the future price distribution. The difference between the upper and the lower bound on costs then has a clear interpretation as costs of limited adjustment.

The results obtained for the example under study are summarized in Table 9-1. It turns out that the optimal capacity choice in period 1 comprises considerably more capacities in combined cycle plants in both settings than in the static long-term equilibrium of period 1, 43.4 resp. 42.6 GW as

252 GW in the static case. This reflects clearly the potential

compared to 38.5 GW in the static case. This reflects clearly the potential usefulness of combined cycle plants in the low gas price scenario. Conversely, the coal capacities are with 31.9 resp. 32.2 GW lower than the 37.5 GW in the static case. The anticipation at stage 0 of what will happen at stage 2 clearly makes not that much a difference on the results. Much more important are the values of key parameters like gas price volatility, interest rate and gas price level. Their impact is illustrated through the parameter variations shown in Figure 9-6 to Figure 9-8.

Table 9-1. Results for two stage stochastic investment problem – Reference case

	Unit	Installed capacity			Investment		
		Period 1	Period 2, high	Period 2 low	Period 1	Period 2, high	Period 2 low
Static expectations							
Gas turbine	GW	3.9	2.6	7.9			
Combined cycle plant	GW	38.5	14.5	72.1			
Coal plant	GW	37.5	62.9	0.0			
With anticipation, full adaptation in period 1							
Gas turbine	GW	4.7	3.3	3.3	4.7	0.0	0.0
Combined cycle plant	GW	43.4	31.0	53.8	43.4	0.0	22.9
Coal plant	GW	31.9	50.0	22.8	31.9	27.2	0.0
Total cost[a]	Mio. €				147.6		
With anticipation, partial adaptation in period 1							
			of which from Period 0				
Gas turbine	GW	5.2 *3.4*	4.1	4.1	1.8	0.0	0.0
Combined cycle plant	GW	42.6 *33.0*	31.6	54.5	9.6	0.0	22.9
Coal plant	GW	32.2 *32.2*	50.0	21.4	0.0	28.6	0.0
Total cost	Mio. €				147.9		

[a] corrected for investments attributable to period 0

For low gas price volatility, the attractiveness of coal fired plants increases, since this capital intensive technology is then less likely to become an inefficient technology in period 2 (cf. Figure 9-6). But from a volatility of 30 % onwards, the share of coal fired power plants remains constant in the setting with partial adjustment. This is due to capacity built up under myopic expectations in period 0 and which is not scrapped in period 1. Yet no new coal capacity is built in period 1 in these cases. If the full capacity can be adjusted in period 1, the share of gas fired plants is increased at higher volatilities since the risk in period 2 is comparatively lower for the gas-fired plants given that these are less capital intensive than the coal-fired plants.

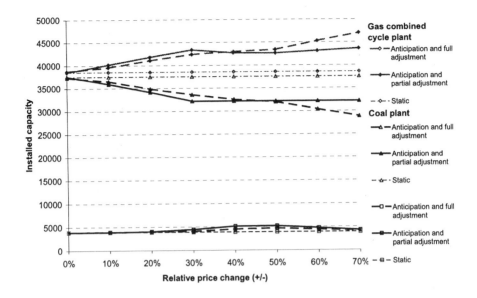

Figure 9-6. Impact of variations in price volatility on installed capacity at the first stage in the two stage decision example

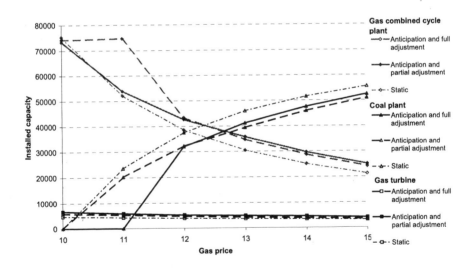

Figure 9-7. Impact of variations in average gas price on installed capacity at the first stage in the two stage decision example

If the average gas price is decreased, the gas technology may become efficient in period 1 even for base load operation (cf. Figure 9-7). This is the case for a gas price of 10 €/MWh, where the share of coal fired plants is zero independently of the methodology used. Here coal plants are only installed in period 2. But in the case with full flexibility at stage 1, no coal plants are installed either at a price of 11 €/MWh, since they might incur heavy losses in period 2. At higher prices, the share of coal plants gradually increases and the share of gas-fired plants decreases. But the combined cycle plant still remains an economically viable alternative for medium load factors, thus it does not disappear completely.

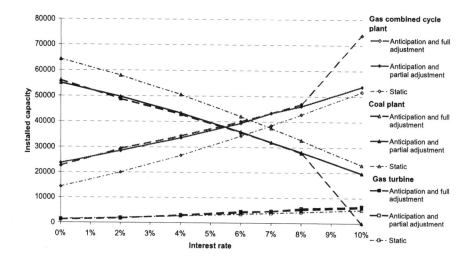

Figure 9-8. Impact of variations in interest rate on installed capacity at the first stage in the two stage decision example

Besides the gas price, the interest rate also exerts a considerable impact on technology choice. As shown in Figure 9-8, a 10 % interest rate will lead to a complete abandonment of the coal technology in period 1, if full flexibility is given. By contrast, at zero interest rate, about 70 % of the installed capacity is in coal-fired plants. For the further analysis one should be aware that by dealing explicitly with price risks, one major source of uncertainty is already covered and thus the rationale for using high discount rates is not that convincing. The results given in Figure 9-8 also point at the fact that lumping together all risks and tackling them through an appropriate choice of the discount rate is not likely to provide the same results as a separate treatment of price risks. This is notably true since the price risks affect the different technologies in a rather different way.

4.3 General solution approach

Of course a possible generalization of the approach outlined in the previous section to a multiple stage setting is to formulate a stochastic program encompassing all the future time steps and possible scenarios. However due to the "curse of dimensionality" this approach may rapidly lead to very large and unsolvable problems. An alternative is to decompose the problem, either exactly or using an approximation. Here an approximate decomposition is proposed which allows obtaining accurate and interpretable results without the computational effort required to reach a consistent overall solution.

The basic idea is to work backward through the scenario tree as in option valuation approaches and in dynamic programming. For each node a two stage optimization problem is formulated, so that the uncertainties of the next planning period are explicitly taken into account (cf. also chapter 6, section 5). The question is then how to cope with the uncertainties at the further decision stages in an at least approximate way. Before giving a formal solution to that problem, a non-formal interpretation is sketched. In fact the decision variables of interest are the capacities to be built at the first stage. The choice of technology at this stage will depend on the fuel prices and other uncertain factors observed at the first, second and further stages, since the earnings accruing from the use of these technologies are distributed over the whole lifetime of the power plant, and the overall cost minimization implies choosing the technology portfolio with the lowest (discounted) lifetime costs. In fact, according to the no-excess profit condition for competitive and efficient markets in neo-classical models (cf. e.g. Shoven, Walley 1992, Böhringer 1996), the investment costs of each technology will in the optimum exactly be covered by the (expected) sum of all operating margins, i.e. differences between prices and operating costs, over the lifetime. The operating margins in the first and second stage are included in a two-stage optimization procedure but what is missing is information on the expected operation margins at the third and subsequent stages. These have to be obtained from the solutions of the two stage problems at the next stage (cf. Figure 9-9).

In fact the expected operating margin will depend on the previously built up stock of the technologies considered. As illustrated in the beginning of section 4 (cf. Figure 9-4), the operating margin is lowered for an inefficient technology and it is also reduced for a technology of which more than the optimal share is installed.

Formally, the operation margin is introduced through a primal decomposition of the optimization problem. The dynamic Eq. (9-35) is first extended to cover explicitly two stages:

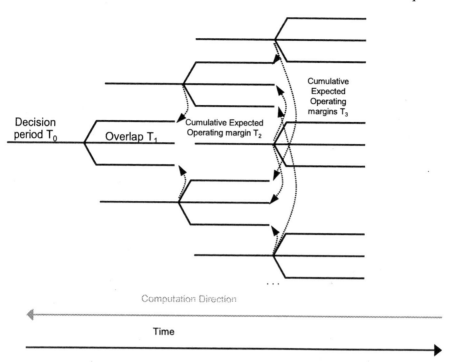

Figure 9-9. Backward induction for investment decisions

$$C_{T,s}\left(\mathbf{K}_{\mathbf{PL},T,s},\mathbf{y},\mathbf{y}_{tot};\widetilde{\mathbf{K}}_{\mathbf{PL},T-1,s_0}\right)=b(\rho,1,d_T)\sum_t\sum_u c_{Op,u,1,s}y_{u,T,s,t}+\sum_u c_{Inv,u}K_{T,s,u}+$$

$$b(\rho,d_T,d_T)\sum_{s'}\mathrm{Pr}_{s\to s'}\Bigg(b(\rho,1,d_{T+1})\sum_{t'}\sum_u c_{Op,u,1,s'}y_{u,T+1,s',t'}+\sum_u c_{Inv,u}K_{T+1,s',u}+$$

$$\sum_{s''}\mathrm{Pr}_{s'\to s''}C_{T+2,s'}{}^*\left(\widetilde{\mathbf{K}}_{\mathbf{PL},T+1,s'}\right)\Bigg)$$

$$\therefore\quad y_{u,T,s,t}\le\widetilde{K}_{\mathbf{PL},T,s,u}$$

$$y_{tot,T,s,t}\le\sum_u y_{u,T,s,t}$$

$$\widetilde{K}_{\mathbf{PL},T,s,u}\le\widetilde{K}_{\mathbf{PL},T-1,s_0,u}+K_{T,s,u}-\zeta_{T-1\to T}\left(\widetilde{K}_{\mathbf{PL},T-1,s_0,u}\right)$$

$$y_{u,T+1,s',t'}\le\widetilde{K}_{\mathbf{PL},T+1,s',u}$$

$$y_{tot,T+1,s',t'}\le\sum_u y_{u,T+1,s',t'}$$

$$\widetilde{K}_{\mathbf{PL},T+1,s',u}\le\widetilde{K}_{\mathbf{PL},T,s,u}+K_{T+1,s',u}-\zeta_{T\to T+1}'\left(\widetilde{K}_{\mathbf{PL},T,s,u}\right)$$

$$(9\text{-}40)$$

The cost function $C_{T+2,s'}{}^*\!\left(\widetilde{\mathbf{K}}_{\mathrm{PL},T+1,s'}\right)$ for the stages $T+2$ and onwards is in turn the outcome of an optimization involving a large number of further decision variables. But for the decisions under study it is sufficient to consider its dependency from the capacity variables $\widetilde{\mathbf{K}}_{\mathrm{PL},T+1,s'}$. Therefore instead of the cost function $C_{T+2,s'}{}^*\!\left(\widetilde{\mathbf{K}}_{\mathrm{PL},T+1,s'}\right)$ the operating margin $\Theta_{T+2,s'}{}^*$ is used which has opposite sign and does not include any costs related to periods beyond $T+1$. The objective function can then be written:

$$C_{T,s}\!\left(\mathbf{K}_{\mathrm{PL},T,s},\mathbf{y},\mathbf{y}_{tot};\widetilde{\mathbf{K}}_{\mathrm{PL},T-1,s_0}\right)=b(\rho,1,d_T)\sum_t\sum_u c_{Op,u,1,s}\,y_{u,T,s,t}\;+$$

$$\sum_u c_{Inv,u}K_{T,s,u}+b(\rho,d_T,d_T)\sum_{s'}\mathrm{Pr}_{s\to s'}\!\left(b(\rho,1,d_{T+1})\sum_{t'}\sum_u c_{Op,u,1,s'}y_{u,T+1,s',t'}\;+\right.$$

$$\left.\sum_u c_{Inv,u}K_{T+1,s',u}-\Theta_{T+2,s'}{}^*\!\left(\widetilde{\mathbf{K}}_{\mathrm{PL},T+1,s'}\right)\right)$$

(9-41)

$\Theta_{T+2,s'}{}^*$ is the value of investments undertaken in $T+1$ (or earlier) during period $T+2$ and later. It is dependent on the initially installed capacities $\widetilde{\mathbf{K}}_{\mathrm{PL},T+1,s'}$ in $T+1$. Since it is itself the result of an optimization calculus it can be shown to be concave in $\widetilde{\mathbf{K}}_{\mathrm{PL},T+1,s'}$. For any other $\widetilde{\mathbf{K}}_{\mathrm{PL},T+1,s'}{}'$ the following inequality thus holds:

$$\Theta_{T+2,s'}{}^*\!\left(\widetilde{\mathbf{K}}_{\mathrm{PL},T+1,s'}{}'\right)\le\Theta_{T+2,s'}{}^*\!\left(\widetilde{\mathbf{K}}_{\mathrm{PL},T+1,s'}\right)+\psi_{\mathrm{PL},T+1,s'}\!\left(\widetilde{\mathbf{K}}_{\mathrm{PL},T+1,s'}{}'-\widetilde{\mathbf{K}}_{\mathrm{PL},T+1,s'}\right)$$
(9-42)

Thereby $\psi_{PL,\ T+1,s'}$ is the (vector of) Lagrangian multipliers of the capacities restrictions involving $\widetilde{\mathbf{K}}_{\mathrm{PL},T+1,s'}$ in the sub-problem for $\Theta_{T+2,s'}{}^*\!\left(\widetilde{\mathbf{K}}_{\mathrm{PL},T+1,s'}\right)$. The inequality (9-42) is hence an approximation to the true function $\Theta_{T+2,s'}{}^*\!\left(\widetilde{\mathbf{K}}_{\mathrm{PL},T+1,s'}\right)$ and combining several such inequalities will yield a concave hull for the function $\Theta_{T+2,s'}{}^*\!\left(\widetilde{\mathbf{K}}_{\mathrm{PL},T+1,s'}\right)$ (cf. Figure 9-10). Combining such inequalities with the objective function (9-41) corresponds to a Bender's decomposition (cf. Benders 1962, e.g. also Nemhauser, Wolsey 1988). By solving the sub-problem on $\Theta_{T+2,s'}{}^*$ for several starting points $\widetilde{\mathbf{K}}_{\mathrm{PL},T+1,s'}$, the true function may be approximated by a series of so-called cuts (the straight lines in Figure 9-10). These cuts constitute an upper bound to the true value of the operational margin as a function of the initial capacities $\tilde{\mathbf{K}}_{PL,T+1,s'}$.

Benders' decomposition has been applied to electricity planning problems in the short-term notably by Pereira and Pinto (1991), but also by Jacobs et al. (1995) Morton (1996) and Takriti et al. (2000). The stochastic dual dynamic programming approach developed by Pereira and Pinto (1991) precisely uses a series of nested Benders cuts to approximate the water value in hydropower scheduling.

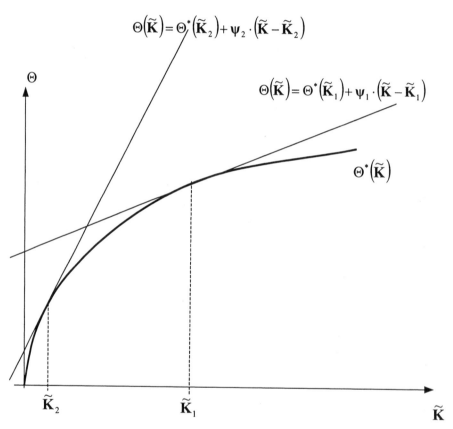

$$\Theta\left(\tilde{\mathbf{K}}\right) = \Theta^*\left(\tilde{\mathbf{K}}_2\right) + \psi_2 \cdot \left(\tilde{\mathbf{K}} - \tilde{\mathbf{K}}_2\right)$$

$$\Theta\left(\tilde{\mathbf{K}}\right) = \Theta^*\left(\tilde{\mathbf{K}}_1\right) + \psi_1 \cdot \left(\tilde{\mathbf{K}} - \tilde{\mathbf{K}}_1\right)$$

$$\Theta^*\left(\tilde{\mathbf{K}}\right)$$

Figure 9-10. Approximation of the operational margin θ through a series of Benders cuts for different initial endowments $\tilde{\mathbf{K}}$

As Pereira and Pinto do, the proposed approach also does not aim at determining the exact solution to the multi-stage problem, but only an approximate solution achievable through backward induction. But in contrast to the approach of Pereira and Pinto, it is proposed here to use a two-stage stochastic program to determine the expected operational margin.

By modeling two stages explicitly the approximation obtained for the cutting plane should be much better than if Monte-Carlo simulations are used to identify initial values for computing the cutting planes.

But still several choices are possible for the initial capacities at the beginning of each two-stage planning problem. Here two obvious choices are considered as in the two-stage example:

No initial capacities except those already existing at the initial stage T_0 and not scrapped up to $T_{k,s}$: in this case the capacities are chosen optimally given the information available at the first decision stage $T_{k,s}$. The system costs obtained in this setting are a lower bound to the overall system costs since any earlier decision deviating from the optimal decision with information available at $T_{k,s}$ will cause an increase in costs for stage $T_{k,s}$ and possibly for subsequent stages.

Preexisting capacities chosen at stage $T_{k-1,s}$ under myopic expectations: this will provide an upper bound to the expected system costs since it is a simple decision rule which takes into account current information without attempting to anticipate beyond current price. It does not take into account the information available at time $T_{k,s}$ or a priori expectations but it is the minimum level of analysis which each decision maker will usually perform.

For the uncertain variables (prices and possibly loads or political decisions), a lattice of possible scenarios is built as described in chapter 6, section 5. Then the two stage optimization problem is solved for each node in the lattice, starting from the last period considered. For the last period, as described in section 4.2, a deterministic, static price scenario is assumed. For the preceding time steps, the two-stage model is solved using the Benders cuts from the following time step to approximate the operating margin obtainable at the subsequent stages.

An advantage of the approach developed is that it may be adapted to different ways of modeling the operation within one year and the resulting operation margin. Simpler models than the peak-load pricing model can be used within one year but also more complex approaches. Notably the fundamental linear programming model developed in chapter 4 to describe the short-term operation can be used to simulate the plant operation within each period. This allows notably including the effect of hydro storage scheduling, start-up costs or other operation restrictions.

5. APPLICATION

The developed methodology is applied to the case of investments in the German electricity market. In the following, the key characteristics of the system investigated and the data used are discussed in section 5.1, then the results obtained are discussed in section 5.2.

5.1   Case study analyzed

In order to cope adequately with the characteristics of the power plants installed in Germany, the fundamental short and medium term model described in section 4.3 is taken as basis for modeling the system operation. Hence, demand fluctuations and plant operation are modeled in the same way as they are in the shorter term horizon. This allows for considerable synergies notably with respect to data collection and handling but also for the model equations.

The load duration curve is represented as before by 144 time segments, with each one representing at most 90 hours. The dynamic constraints linking different hours are taken into account in this approach, allowing to model start-up costs, hydro storage and minimum operation and shut-down times as explained in section 4.1 and section 4.3.2. In order to limit the size of the model, the interconnections with the European neighboring countries are however neglected in the current version of the long-term model.

Furthermore the focus is on modeling the impact of price uncertainties on the optimal investment choice and therefore electricity demand is assumed to remain constant over the planning horizon (In Germany, it has increased by less than 1 % p. a. on average during the last decade, cf. chapter 3). Also technological and political uncertainties, which are currently of considerable importance, are not directly included into the model. For modeling fuel price uncertainty, the fuel price model developed in section 4.3.1 is taken as basis. 1000 simulation runs are carried out which are then aggregated to 15 price scenarios per time step for the stochastic model. Since the estimated fuel model focuses on the changes of (short-term) price changes as explained variable it will however not necessarily provide good long-term forecasts. Rather fundamental analyses should be used to estimate the most likely longer term evolution of fuel prices. Such analyses have been carried out notably for the European Commission using the POLES model (cf. Criqui 1996). But also for the Enquete Commission of the German Bundestag on "Sustainable Energy Supply in the Context of Liberalization and Globalization" such likely price scenarios have been identified (cf. Enquete Commission 2002, Fraktionen von CDU/CSU und FDP 2002). These price scenarios are used here to define the general development, whereas the fluctuations around this trend are taken from the stochastic model. Given that the stochastic model is formulated in logs of prices, the mean logarithmic price path is calibrated to the pre-specified scenario. The corresponding development of the median energy prices is plotted in Figure

9-11 for the most important energy carriers. Gas prices are hence expected to decrease relatively to their year 2000 level until 2005, but afterwards prices are projected to increase at an average annual rate of 1.5 %. For the other fossil fuels rising prices are also expected as a consequence of increasing resource scarcity, notably of oil and gas. The price of nuclear fuels is by contrast taken as constant on average[62].

The other energy carriers by contrast are more stable and exhibit mean reversion. Given the extreme price spikes simulated for gas prices, the mean value is also considerably bended upwards, whereas the median value, i.e. the value not exceeded in more than 50 % of the simulated price paths is growing much less.

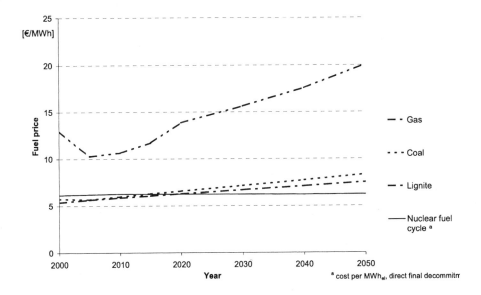

Figure 9-11. Median price paths for energy carriers for the stochastic investment model

The observed gas price development until September 2002 is also shown in Figure 9-12, indicating that so far the optimistic expectations on declining gas prices have not been fully met by the reality.

---

[62] Note that investments in nuclear energy have, contrarily to the current policy of the German government, not been excluded a priori in the following analysis. Of course, investments in nuclear power plants require a detailed analysis of the technical and environmental risks associated with this technology, which is beyond the scope of this book. In a comparative analysis then obviously also the risks of other technologies (e.g. climate change) have to be assessed.

For investment in electricity generation, the key alternatives summarized in Table 9-2 are considered. Distributed and renewable technologies are not included here since it is expected that they will only contribute substantially to electricity generation in the next decade if they are supported and subsidized through specific political instruments. The introduction and modification of the corresponding political instruments clearly contributes to increasing the uncertainty for conventional electricity generation investments, by notably increasing the uncertainty on the load to be covered by conventional plants. But for the promoted technologies the uncertainty is usually reduced, in the case of fixed feed-in tariffs (as currently applied e.g. in Germany, France and Spain) the market price uncertainty is even fully eliminated[63].

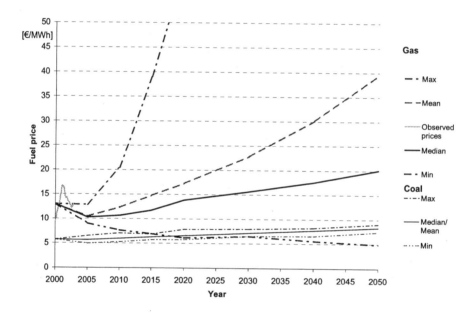

Figure 9-12. Upper and lower bounds, median and mean values of price paths for gas and coal

As shown already in section 4.2, the choice of the interest rate has a key impact on investment valuation. Also in recent applications to energy investment issues a broad range of interest rates has been applied. Grobbel (1999), following Bunn et al. (1997) uses for example a real interest rate of

---

[63] Whether such a risk elimination (or rather risk transfer) provides adequate incentives for investments has to be analyzed in some detail, but such an analysis is beyond the scope of the present book.

12 % p.a.. Frometa et al. (2000) applies a nominal interest rate of 11 % and real interest rate of 8 % for valuing power plant investments whereas the scenario calculations for the aforementioned Enquete commission have been done using a (real) interest rate of 4 %. Even a real interest rate of 3.5 % has been recently used to evaluate climate change policies in Germany (cf. Fahl et al. 2003).

Table 9-2. Technical and economic characteristics of the power plants considered for investment

		Coal fired plant with atmospheric dust combustion	Lignite fired plant with atmospheric dust combustion	Gas turbine	Gas fired combined cycle plant	Pressurized water reactor, advanced DWR-1400 concept
Net power output	MW	750	750	146	750	1375
Overall fuel efficiency	%	43	40	33	57.6	34
Planning and construction time	a	5	2	2	4	7
Economic lifetime	a	35	35	35	35	40
Duration of decommitment and deconstruction	a	0	0	0	0	13
Investment costs including site preparation costs and interest payments during construction phase	€/kW	1050	1350	230	450	1800
Decommissioning and deconstruction cost	€/kW	38	38	38	38	259
Fixed operation costs	€/kW	44	13	11	20	22
Variable costs besides fuel costs	€/MWh	2.2	1.7	1.2	1.2	0.5

For an appropriate choice of the interest rate, the various risks associated with power plant investments have to be analyzed and if possible corresponding market risk prices have to be identified. Notably the following elements will influence the selection of an adequate discounting factor:

- market interest rate for "risk free" assets,
- credit risk (counterparty risk) associated with the investor and corresponding mark-ups on interest rates for loans,
- share of equity (own capital) required for the investment and the remuneration of this capital,
- technical, financial and other project risks,
- market price risks,
- market quantity risks.

Given the current morose economic situation, the market interest rate for short-term loans from central banks is at historical lows, in the Euro zone it has been fixed to 2 % (in nominal terms) since June 6, 2003, whereas in preceding years it turned around 2.5 % to 5 %. But for power plant investments the short-term interest rate certainly is not the correct reference value, rather the interest paid on long-term government bonds should be taken as a reference. This is currently in Germany at about 5 % p.a. for a 10-year bond (cf. Deutsche Bundesbank 2003). Since more than one year annual inflation has been at around 1 % p. a. in Germany (cf. Statistisches Bundesamt 2003) and no major changes are anticipated, so that the corresponding real interest rate is around 4 % p.a.

The credit risk associated with utilities has lead to a mark-up of between 0.75 % and 1 % p.a. on recent obligations emitted by German utilities RWE and EnBW (Leuschner 2001, Leuschner 2002) compared to governmental bonds. However this is the risk associated with vertically integrated utilities which have also important assets in the regulated grid business. Thus for the valuation of power plant investments in a competitive market environment rather the upper bound should be used, leading to a company debt interest rate of 5 % p.a. in real terms.

A conventional estimate on own capital requirements for companies in Germany is 20 % (equity-to-assets ratio). Among the large utilities, E.On and Vattenfall Europe are currently the only ones to exceed this share slightly with 22 % resp. 21 % (cf. EnBW 2003, E.On 2003, RWE 2003, Vattenfall Europe 2003). Yet for the risky power plant investments such a share may well be required also by borrowers (cf. Frometa et al. 2000). In line with the CAPM approach (cf. e.g. Ross 1977), this capital should be remunerated with an excess return depending on the market price of risk and the correlation between general market risk and company risk. Unfortunately, no robust results are available for such a valuation and a rather broad range of empirical results may be expected as shown by Hartung (2002) for the case of insurance companies. But as a first approximation, a 10 % return on equity in real terms seems adequate. According to the weighted capital cost approach (WCCA), this leads to an average capital cost of 6 % p. a. in real terms[64].

These capital costs still do not include risk premiums for project specific risks, notably that technical and economic targets are met during construction and operation and that no premature failures occur. Partly these risks are covered by insurance premiums paid by the power plant operator

---

[64] RWE (2003) uses a hypothetical equity-to-assets ratio of 40 % to derive a nominal WACC of 10 % for its electricity division.

and the construction company (which are included in the fixed operation costs), but still some risks have to be born by the operator himself. Here an average risk premium of 1 % p. a. is estimated, with certainly a larger part of the risk occurring at the front end during the construction and commissioning phase[65]. This leads to an overall real discount rate of 7 % p. a. to be applied to power plant investments, or 8 % in nominal terms at current inflation rates in Germany.

Market price risks are dealt with explicitly in the approach developed here and should therefore not be incorporated in the discount rate. The quantity risk for annual electricity demand is also limited, given that electricity is nowadays at least in industrial countries a basic good with few substitutes. As indicated in section 3.2, demand growth rates have fluctuated in the last decade but have only been negative under the specific circumstances of German reunification. Also they never exceeded 7 %, so that the longer term quantity risks on the demand side seem manageable. More substantial are the risks on the supply side arising from additional generation built under the renewable energy law (EEG). But this politically induced uncertainty shall not be considered in the following, since it is rather difficult to quantify.

5.2   Results

For the described system, the results depicted in Figure 9-13 are obtained for the year 2005. Obviously, the optimal investment decision for most scenarios is not to invest at all. Only in the case of low gas prices, investment in new gas-fired plants is undertaken. This is especially true for scenario 3, where at the same time the coal prices are rather high. To a lesser extent, this applies also to scenario 6.

The installed new capacities vary broadly between scenarios, indicating that partly existing plants, which have not come to the end of their technical lifetime, are replaced because they are no longer part of the efficiency line. One might notice that at the price level of October 2002 (about 12 €/MWh), no investment into new generation would occur according to the model and that gas power plants will only be considered, if the fuel prices are indeed decreasing until 2005 as expected by the Enquete Commission (2002).

---

[65] This approximation of technical risks is obviously rather crude and should be detailed in further analyses. Especially for nuclear energy, the results of numerous risk analyses (E.g. GRS 1979) have to be taken into account and the ethical and social acceptability of investments has to be considered in addition to market-oriented analyses like those presented here.

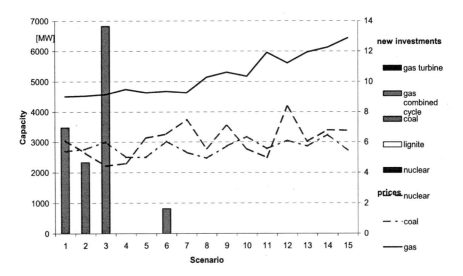

Figure 9-13. Development of investment and fuel prices in different scenarios in 2005

Figure 9-14 gives the corresponding results for the year 2010. Here investment occurs in all scenarios, albeit to a varying degree. In scenario 2 with low gas prices and high other fuel prices, large investments occur again, leading to a replacement of existing power plants by gas combined cycle plants. Investment in gas turbines hardly occurs in these scenarios, indicating that the current number of peaking units still will be sufficient in 2010. Also coal and lignite plants are not chosen for investment, independently of the scenario. By contrast, investment in nuclear power plants is the preferred investment alternative in five scenarios with a total probability of 29 %.

The development in the subsequent years is shown in a more compact form in Figure 9-15. Here the probability-weighted averages of the optimal stock of new power plants in the different scenarios are indicated. Up to the year 2020, gas combined cycle plants dominate among the newly built plants, but under the specified price scenarios the investment shifts after 2020 gradually towards nuclear power plants. Coal and lignite plants do not appear as part of the efficient solution for the given data set.

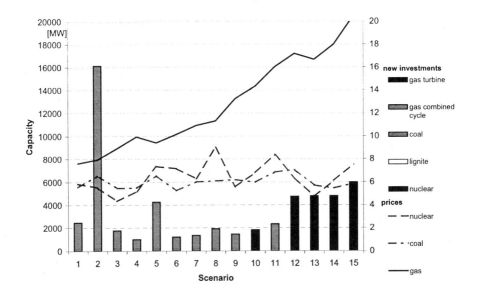

Figure 9-14. Development of investment and fuel prices in different scenarios in 2010

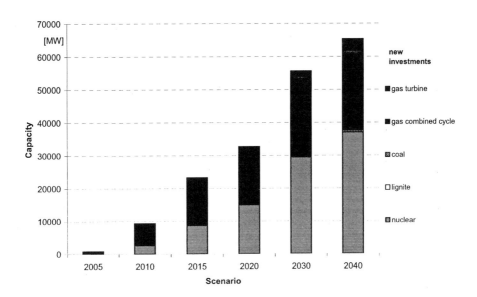

Figure 9-15. Development of the optimal stock of new power plants between 2005 and 2040 averaged over scenarios

Since there is no unique development path, the actual development is in fact not predictable. Also the probability-weighted average is mainly intended to give an idea on the direction of the development "in the mean". But as the longer term decisions need not to be taken today, it is also not necessary to provide unique answers at this stage – instead it is important to value flexibility correctly in such a decision setting under uncertainty. That the uncertain future developments have an impact on the decisions is illustrated through Figure 9-16. Here the optimal stock of new power plants is shown, which is obtained if decision making under certainty is assumed. Thereby the price development corresponds to the mean of the uncertain price scenarios. Taking away the uncertainty shifts the investment towards capital-intensive nuclear technologies from 2015 onwards. In 2010 some investment is done in gas combined cycle plants, but the investment level is lower than in the baseline case. This is because the model anticipates the increasing gas prices, which will reduce the operation margin of combined cycle plants in later years.

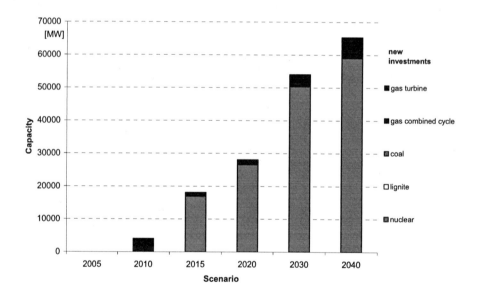

Figure 9-16. Development of the optimal stock of new power plants between 2005 and 2040 under certainty

At first sight, this result contrasts with the findings of Dixit and Pindyck (1994), who derive for simple decision settings that uncertainty will lead to a postponement of investments. But this apparent contradiction can be resolved by observing that Dixit and Pindyck mainly deal with the decision to invest or not into one single technology, which moreover is not facing an increasing cost trend. The general result of Dixit and Pindyck and others that uncertainty will push towards flexible, less-capital intensive technologies is yet clearly confirmed in the present case. But as shown in Figure 9-13 to Figure 9-15 this does not preclude investment in capital-intensive technologies, it only sets higher profitability targets.

One may be surprised that the technologies with the highest fuel price risk, namely gas technologies, are chosen at all in this setting, since they include a substantial danger of being no longer efficient if prices change. But the fuel price risk is at the same time an opportunity for the gas technologies and a thread to all others: the nuclear technology is also indirectly affected by the gas price uncertainty. If gas prices drop, then in turn the nuclear technology becomes inefficient and has to struggle with a reduced operation margin. This is a consequence of the endogenous price setting, which occurs within the developed model. The resulting danger of being not part of the efficiency line is even more problematic for the capital-intensive technologies since these have to recover higher capital costs.

The effect of a phase-out on nuclear energy as currently in place in Germany is illustrated in Figure 9-17. The investment shifts in this case towards coal power plants, but the proportion of gas fired plants is higher than in the baseline case. Since in the baseline case no investment in nuclear power is optimal in 2005, the cost function for 2005 is unchanged in the model. But obviously for later years higher costs have to be incurred. This implies – all precautions taken for the incomplete model specification (e.g. no demand growth) – that the decision on a moratorium for new nuclear power plants is not costly if it is revised before the investments for 2010 are decided.

In this way also other political decisions can be analyzed, determining whether or to what extent they are causing extra costs in the system and when these extra costs do occur. Another alternative for dealing with these uncertainties is to incorporate them directly in the scenario tree. However this requires the use of subjective – or at best intersubjectively agreed – probabilities for the scenario construction.

In general, the results obtained with this new approach of backward induction of dynamic peak-load pricing equilibria are promising but further details have to be implemented in the methodology before it can provide reliable results on what are adequate investment strategies in the future. Notably, short-term load uncertainty longer term demand growth need to be described in detail to provide a more adequate picture of operation margins and corresponding investment opportunities in the liberalized electricity markets.

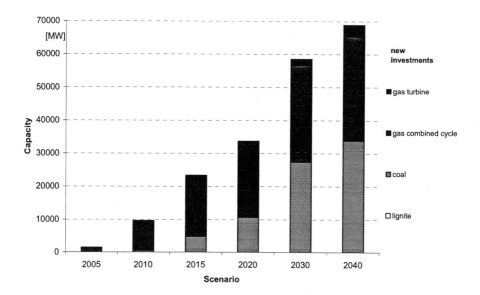

Figure 9-17. Development of the optimal stock of new power plants between 2005 and 2040 without investments in nuclear energy

Chapter 10

FINAL REMARKS

The analyses carried out in the previous sections have highlighted the key role of uncertainties for the planning process in deregulated electricity markets.

The integrated market simulation model developed in chapter 4 has demonstrated how a combination of fundamental analysis and stochastic modeling can be used for price simulation in order to cope simultaneously with the technical restrictions governing electricity production and transmission and with the stochastics observed in the system. Also this approach models the dependency of electricity prices on fuel prices and it can be used to derive both price forecasts and uncertainty ranges for the future development of prices. The numerical results obtained show that the fundamental model explains quite well the average prices observed outside the peak hours and the combination of the fundamental and the stochastic model provides both reasonable price forecasts and plausible volatility estimates These can be used both for the operational management of generation and trading portfolios and for assessing the risks associated with these portfolios.

Still there is potential for further improvement of this modeling concept. Notably a modeling of regime shifts, corresponding to situations with and without capacity constraints is expected to improve the modeling quality further. Additionally, a more detailed treatment of market power, possibly in combination with a game theoretic approach would be of interest. Furthermore, it should be analyzed how the model could be extended to provide also consistent estimates of forward price volatility.

In contrast to many other game theoretic models to describe the electricity markets, the model described in chapter 5 focuses on the retail competition of vertically integrated companies. It thus allows assessing the importance of various factors on the competitive situation of combined generation and retail companies. The numerical results obtained indicate that both competitive cost structures and a sound market basis with loyal customers are key factors for obtaining positive commercial results. It turns also out that substantial extra profits for the suppliers are still possible even in a competitive environment, especially in customer segments like households with low price sensitivity. It has however been shown both theoretically and numerically that this market segment is not necessarily an attractive field for expansion plans, since customer switching can only be obtained at high costs and competitors may well counteract acquisition efforts through own price reductions.

As with many game theoretic models, the essential results are more qualitative than quantitative and clearly further investigations are desirable to obtain reliable estimates on the price sensitivity of the customers. Then the tool could also become a worthy tool for quantitative analysis. On the other hand, the derived global regularity properties make the basic model also an attractive candidate for further refinement and extension, e.g. for the inclusion of pure retail companies as separate players and for an analysis of the interaction between wholesales and retail markets.

Chapter 6 has shown that operation planning in an incomplete market structure such as the continental European one requires specific methods and tools which take into account both the possibilities and the limitations of the given market structure. Again the combination of traditional engineering approaches for unit commitment and load dispatch with finance methods for option valuation have turned out to provide appropriate and workable approaches for solving the short-term operation planning problems. A particular focus has thereby been laid on CHP systems, which even in liberalized electricity markets require a simultaneous optimization of several units given that heat cannot be purchased or sold on a short-term market. For the long-term portfolio management also an approach has been developed which allows determining the optimal hedge for a CHP system. The numerical results indicate that the value of both conventional and CHP plants is substantially higher under price uncertainty than determined through deterministic planning models. This option value is particularly important for plants with high variable costs, such as gas-fired units, which will hardly operate at average prices but are an interesting option for coping with high prices.

Future developments in this field should notably aim at developing efficient decomposition approaches, which allow dealing with a broad range of price and quantity uncertainty in reasonable computation time. Furthermore the developed methods could be easily extended to distributed generation systems, taking into account not only temporal but also spatial variations in supply and demand. In this context particular emphasis should be laid on a combination of stochastic modeling of quantity risks (e.g. from wind production) with stochastic optimization of power plant operation.

Given the high volatility for spot market electricity prices, risk management plays a crucial role in the electricity industry. But so far mostly approaches from the finance sector have been transposed to the electricity market without sufficiently reflecting the specificities of this market. Notably, a large number of products are traded on the forward and future markets. But often these products have only limited liquidity and moreover they do not cover fully the price risks of the underlying spot market products. To cope with this situation, the concept of integral earnings at risk (IEaR) has been developed, which includes both short-term trading risks and the longer-term risks at delivery. It boldly corresponds to a combination of value at risk (VaR) and profit-at-risk (PaR) methods that specifically accounts for the possibilities and impossibilities for risk management in electricity markets. The approach is applied to various power plant portfolios and it is shown that hedging allows reducing substantially the short-term risk but that the delivery risk remains substantial and could only be reduced if the market for shorter-term forwards would become more liquid.

The developed approach offers a huge potential for application to the diverse types of portfolios present in the electricity industry. In particular the combination of price and quantity risks present in hydro storage power plants or full service end user contracts should be analyzed in detail. Also the development of accurate but time saving approximations for risk assessment purposes is certainly a worthwhile task.

The technological uncertainties in the longer run are discussed in chapter 8, focusing on the fuel cell technology which is often perceived as a key technology for the energy systems of the future. A combination of several approaches, including technology analysis, use of experience curves, energy demand analysis and operation simulation, allows assessing the potentials of this emerging technology with much more detail. Obviously the uncertainty cannot be fully removed but the analyses show that under favorable energy price conditions fuel cells become an attractive option for investment in industrial cogeneration at a price level of about 1,000 €/kW$_{el}$. Under less favorable conditions (high gas and low electricity prices) investment cost have to be below 600 €/kW$_{el}$ for a wider penetration of fuel cells into the market. At the current stage, fuel cells are certainly not an opportunity for doing large bets for utilities, but they are an alternative, which should be scrutinized carefully, and limited engagements may represent a valuable option for larger future investments.

The methodology as developed and applied in this chapter has certainly specificities related to CHP and fuel cell technologies, but the general lines are also transferable to other new technology lines, such as photovoltaics or biomass-fuelled plants. A topic, which requires further extended investigations along these lines, are distributed energy systems combining fuel cell and other decentralized cogeneration technologies with fluctuating renewables and energy storage. Here detailed analyses on the interaction of new generation and grid technologies with energy and reserve power markets are needed.

The interactions between energy price developments and technology choice are looked at in detail in chapter 9. The approach developed there allows taking into account for fuel price uncertainty in the longer run and proceeds by backward induction from possible long-term equilibria towards today's decisions. It is based on the principle of peak-load pricing but it incorporates also the operation restrictions for power plants, taking up the fundamental model developed in chapter 4. The methodology is applied in a case study to the German market and it is shown that the results obtained differ substantially from those obtained using conventional static or deterministic planning approaches.

The model could be developed further to incorporate also the impact of load (and fluctuating generation) uncertainty on optimal investment. Furthermore, the model can in principle also cope with other types of uncertainties such as technological or political uncertainty. Here future research is needed to clearly identify the potentials and limitations of the developed approach. Additionally investigations are needed on the robustness of the computed investment-price equilibria. I.e. the question has to be addressed, whether under the current market structures investments will be done sufficiently early to prevent power shortages, given the combination of substantial price uncertainty and long lead times required for power plant planning and construction. This is certainly also the key challenge for market design in the future - ensuring supply adequacy on competitive markets at lowest possible costs.

References

AG Energiebilanzen (2001): Auswertungstabellen zur Energiebilanz für die Bundesrepublik Deutschland 1990 bis 2000, Stand 9/2001. Berlin also: http://www.ag-energiebilanzen.de/, accessed August 4, 2003.

Albiger, J. (1998): Integrierte Ressourcenplanung in der Energiewirtschaft mit Ansätzen aus der Kraftwerkseinsatzplanung. Forschungsbericht des IER, Universität Stuttgart Band 43. Stuttgart.

Ali, P. (2000): Weather Derivatives, Hedging Volumetric Risk and Directors' Duties. In: Company and Securities Law Journal 18, pp. 151 - 154.

Anderson, S. P., de Palma, A. (1992): The logit as a model of product differentiation. In: Oxford Economic Papers 44, pp. 51 – 67.

Anderson, S. P., de Palma, A., Nesterov, Y. (1995): Oligopolistic competition and the optimal provision of products. In: Econometrica 63, pp. 1281 – 1301.

Andersson, B., Bergman, L. (1995): Market structure and the price of electricity: An ex ante analysis of the deregulated Swedish electricity market. In: The Energy Journal 16, pp. 97 – 110.

Argote, L., Epple, D. (1990): Learning Curves in Manufacturing. In: Science 247, pp. 920 – 924.

Arrow, K. J. (1962): The Economic implications of Learning by Doing. In: Review of Economic Studies 29, pp. 155 - 173.

Artzner, P., Delbaen, F., Eber, J.M., Heath, D. (1999): Coherent Measures of Risk. In: Mathematical Finance 9, pp. 203 - 228.

Auer, H., Targner, M., Keseric, N., Haas, R. (2002): How to Ensure the Optimal Use of the European Transmission Grid. Presentation at the 25th Annual IAEE International Conference, Aberdeen June 26 – 29, 2002.

Bacaud, L., Lemarechal, C., Renaud, A., Sagastizabal, C. (2001): Bundle methods in stochastic optimal power management: A disaggregated approach using preconditioners. In: Computational Optimization and Applications 20, pp. 227 - 244.

BAFA - Bundesamt für Außenwirtschaft (2002): Energie-Statistiken http://www.bafa.de/1/de/ service/statistik/statistik.htm and http://www.bmwi.de/Redaktion/Inhalte/Downloads /ausgewaehlte - statistiken,templateId=download.pdf, accessed August 4, 2003.

Bagemihl, J. (2002): Optimierung eines Portfolios mit hydro-thermischem Kraftwerkspark im börslichen Strom - und Gasterminmarkt. Forschungsbericht des IER, Universität Stuttgart Band 94. Stuttgart.

Baldick, R., Hogan W. (2001): Capacity Constrained Supply Function Equilibrium Models of Electricity Markets: Stability, Non - decreasing Constraints, and Function Space Iterations. Working papers series of the Program on Workable Energy Regulation (POWER) PWP - 089. Berkeley CA.

Barlow, M. T. (2002): A diffusion model of electricity prices. In: Mathematical Finance 12, pp. 287 - 298.

Barnwell, C. D. (2001): Unmanaged Generation Portfolios - The Risk of Doing Nothing. In: Power & Gas Marketing July/August 2001, pp. 36 - 38.

Baumol, W., Panzar, J., Willig, R. (1982): Contestable markets and the theory of industry structure. San Diego CA.

BCG - Boston Consulting Group (1972): Perspectives on Experience. Boston MA.

BDI – Bundesverband der deutschen Industrie, VIK – Verband der Industriellen Energie - und Kraftwirtschaft, VDEW – Verband der Elektrizitätswirtschaft, VDN - Verband der Netzbetreiber, ARE - Arbeitsgemeinschaft regionaler Energieversorgungs- Unternehmen, VKU - Verband kommunaler Unternehmen (2001): Verbändevereinbarung über Kriterien zur Bestimmung von Netznutzungsentgelten für elektrische Energie und über Prinzipien der Netznutzung vom 13. Dezember 2001. Berlin et al.

BDI – Bundesverband der deutschen Industrie, VIK – Verband der Industriellen Energie - und Kraftwirtschaft, VDEW – Verband der Elektrizitätswirtschaft (1998): Verbändevereinbarung über Kriterien zur Bestimmung von Durchleitungsentgelten vom 22. Mai 1998. Berlin et al.

BDI – Bundesverband der deutschen Industrie, VIK – Verband der Industriellen Energie - und Kraftwirtschaft, VDEW – Verband der Elektrizitätswirtschaft (1999): Verbändevereinbarung über Kriterien zur Bestimmung von Netznutzungsentgelten für elektrische Energie vom 13. Dezember 1999. Berlin et al.

Benders, J. F. (1962): Partitioning procedures for solving mixed - variables programming problems. In: Numerische Mathematik 4, pp. 238–252.

Bergman, L., Brunekreeft, G., Doyle, C., von der Fehr, N. - H., Newbery, D., Pollitt, M., Regibeau, P. (2000): A European Market for Electricity? Monitoring European Deregulation 2nd report. London.

Bergschneider, C., Karasz, M., Schumacher, R. (1999): Risikomanagement im Energiehandel. Grundlagen, Techniken und Absicherungsstrategien für den Einsatz von Derivaten. Stuttgart.

Berthold, O., Bünger, U., Niebauer, P., Schindler, J., Schurig, V., Weindorf, W. (1999): Analyse von Einsatzmöglichkeiten und Rahmenbedingungen von Brennstoffzellen-systemen in Haushalten und im Kleinverbrauch in Deutschland und Berlin. Gutachten im Auftrag des Büros für Technikfolgenabschätzung (TAB). Ludwig - Bölkow - Systemtechnik (LBST). Ottobrunn.

Bhattacharya, K., Bollen, M. H. J., Daalder, J. E. (2001): Operation of Restructured Power Systems. Boston

Birol, F. (2003): World Energy Outlook 2020 and Global Electricity Investment Challenges. Presentation at the conference: Power Generation Investment in Liberalised Electricity Markets, organized by the Nuclear Energy Agency, March 25 – 26, 2003. Paris. Also: http://www.nea.fr/html/ndd/investment/session1/Birol.pdf, accessed August 4, 2003.

Black, F., Scholes, M. (1973): The Pricing of Options and Corporate Liabilities. In: Journal of Political Economy 81, pp. 637 - 659.

Blesl, M. (2002): Räumlich hoch aufgelöste Modellierung leitungsgebundener Energie-versorgungssysteme zur Deckung des Niedertemperaturwärmebedarfs. Forschungsbericht des IER, Universität Stuttgart Band 92. Stuttgart.

Blesl, M., Fahl, U. (2002): Szenarien einer liberalisierten Stromversorgung im Jahr 30230 in Baden-Württemberg. Endbericht IER, Universität Stuttgart.

Bohn, D., Leepers, J., Sürken, N., Krüger, U. (2000): Kopplung von Mikrogasturbinen mit Brennstoffzellen. In: BWK 52 (5), pp. 48 – 51.

Böhringer, C. (1996): Allgemeine Gleichgewichtsmodelle als Instrument der energie- und umweltpolitischen Analyse. Frankfurt.

Boiteux, M. (1960): Peak - load pricing. In: Journal of Business 33, pp. 157 - 179.

Bolle, F. (1992): Supply Function Equilibria and the Danger of Tacit Collusion: The Case of Spot Markets for Electricity. In: Energy Economics 14, pp. 94 - 102

Bolle, F. (2001): Competition with Supply and Demand Functions. In: Energy Economics 23, pp. 253 - 277.

Bollerslev, T. (1986): Generalized Autoregressive Conditional Heteroskedasticity. In: Journal of Econometrics 31, pp. 307 - 327.

Bollerslev, T. (1990): Modeling the coherence in short - run nominal exchange rates: a multivariate generalized Arch model. In: Review of Economics and Statistics 74, pp. 498 - 505.

Borenstein, S., Bushnell, J., Knittel C. R. (1999): Market Power in Electricity Markets: Beyond Concentration Measures. In: The Energy Journal 20, pp. 65 - 88.

Botterud, A., Bhattacharya, A. K., Ilic, M. (2002): Futures and spot prices – an analysis of the Scandinavian electricity market. In: Proceedings 34th Annual North American Power Symposium (NAPS 2002). Tempe.

Box, G. E. P., Cox, D. R. (1964): An analysis of transformations. In: Journal of the Royal Statistical Society B 26, pp. 211 - 252.

Box, G. E. P., Jenkins, G. M., Reindels, G. C. (1994): Time series analysis. Forecasting and control. Upper Saddle River.

Brand, H. (2004): Ein Verfahren zur Bestimmung der Angebotsfunktion für den Spotmarkt im Rahmen der stochastischen, kurzfristigen KWEP unter spezieller Berücksichtigung technischer Restriktionen von KWK - Systemen. Ph. D. under preparation. Stuttgart.

Brand, H., Thorin, E., Weber, C., Madlener, R., Kaufmann, M., Kossmeier, S. (2002): Market analysis and tool for electricity trading. OSCOGEN Deliverable D5.1. http:/www.oscogen.ethz.ch, accessed August 4, 2003.

Brand, H., Weber, C. (2001): Stochastische Optimierung im liberalisierten Energiemarkt - Wechselwirkungen zwischen Kraftwerkseinsatz, Stromhandel und Vertrags-bewirtschaftung. In: VDI (Hrsg.): Optimierung in der Energieversorgung IV. VDI-Berichte Band 1627, Düsseldorf 2001, pp. 173 – 183.

Brand, H., Weber, C. (2002): Report on the optimum bidding strategy derived and performance evaluation. OSCOGEN Deliverable D 5.2. http:/www.oscogen.ethz.ch accessed August 4, 2003.

Braun, E. (1998): Technology Assessment in Context. London, New York.

Breiman, L. (1986): Probability. New York.

Brooke, A., Kendrick, D., Meeraus, A., Raman, R. (1998): GAMS – A User's Guide. Washington, Köln, also: http://www.gams.com/docs/gams/GAMSUsersGuide.pdf, accessed August 4, 2003.

Brouwer, L. E. J., (1910): Über die Abbildung von Mannigfaltigkeiten, Mathematische Annalen 71, pp. 97 - 115.

Brown, G., Johnson, M. B. (1969): Public Utility Pricing and Output under Risk. In: American Economic Review 59, pp. 119 – 128.

Brunekreeft, G. (2002): Regulation and third-party discrimination in the German electricity supply industry. In: European Journal of Law and Economics 13, pp. 203 - 220.

Bunn, D. W., Bower, J., Wattendrup, C. (2001): Model based comparison of strategic consolidation in the German electricity industry. In: Energy Policy 29, pp. 987 - 1005.

Bunn, D. W., Larsen, E. R., Vlahos, K. (1997): Modelling Approaches for Electricity Privatization: Developing Strategies for America's Electric Utilities. In: Bunn, D. W., Larson, E. R. (eds.): Systems Modelling for Energy Policy. Chichester.

Campbell, J. Y., Lo, A. W., MacKinlay, A. C. (1997): The Econometrics of Financial Markets. Princeton NJ.

Caplin, A., Nalebuff, B. (1991): Aggregation and Imperfect Competition. In: Econometrica 51, pp. 25 - 60.

Carlton, D. W. (1977): Peak Load Pricing with Stochastic Demand. In: American Economic Review 67, pp. 1006 – 1010.

Caroe, C. C., Schultz, R. (1999): Dual decomposition in stochastic integer programming. In: Operations Research Letters 24, pp. 37 - 45.

CEC – Commission of the European Communities (2002): First benchmarking report on the implementation of the internal electricity and gas market. Commission Staff Working Paper SEC (2001) 1957, Updated Version with Annexes March 2002. http://europa.eu.int/ comm/energy/electricity/benchmarking/doc/1/report-amended_en.pdf

CEC – Commission of the European Communities (2003): Second benchmarking report on the implementation of the internal electricity and gas market. Commission Staff Working Paper SEC (2003) 448, Updated report incorporating candidate countries. http://europa.eu.int/comm/energy/electricity/benchmarking/doc/2/sec_2003_448_en.pdf

Chao, H.-P. (1983): Peak load pricing and capacity planning with demand and supply uncertainty. In: Bell Journal of Economics 14, pp. 179 – 190.

Chi, P., Russell, C. T. (1999): Statistical Methods for Data Analysis in Space Physics. http://www - ssc.igpp.ucla.edu/personnel/russell/ESS265/Ch9/autoreg/, accessed August 4, 2003.

Claeson, U. (2000): Using the experience curve to analyze the cost of development of the combined cycle gas turbine. In: Wene, C.-O., Voß, A., Fried, T. (eds.): Experience Curves for Policy Making - The Case of Energy Technologies. Forschungsbericht des IER, Universität Stuttgart Band 67, pp. 101 - 121.

Clewlow, L., Strickland, C. (2000): Energy derivatives. Pricing and risk management. London.

Cochrane, J. H. (1989): How Big is the Random Walk in GNP. In: Journal of Political Economy 96, pp. 893 – 920.

Conejo, A.J., Arroyo, J.M., Jimenez Redondo, N., Prieto, F.J. (1999): Lagrangian relaxation applications to electric power operations and planning problems. In: Song, Y.H. (ed.): Modern Optimisation Techniques in Power Systems. Dordrecht, pp. 203 - 228.

Courtney, H., Kirkland, J., Viguerie, P. (1997): Strategy Under Uncertainty. In: Harvard Business Review November - December, pp. 67 - 79.

Courtney, H., Kirkland, J., Viguerie, P. (2001): Strategy under uncertainty. In: McKinsey Quarterly December 2001, pp. 5 – 14.

Cowan, R. (2000): Learning Curves and Technology Policy: On Technology Competitions, Lock.in and Entrenchment. In: Wene, C.-O., Voß, A., Fried, T. (eds.): Experience Curves for Policy Making - The Case of Energy Technologies. Forschungsbericht des IER, Universität Stuttgart Band 67, pp. 33 – 51.

Crampes, C., Creti, A. (2002): Capacity Competition in Electricity Markets. Working paper IDEI, University of Toulouse.

Crew, M.A., Kleindorfer, P. R. (1976): Peak Load Pricing with a Diverse Technology. In: Bell Journal of Economics 7, pp. 207 – 231.

Criqui, P. (1996): International markets and energy prices: the POLES model. In: Lesourd, J.B., Percebois, J., Valette, F. (eds.): Models for energy policy. London, New York, pp. 14 – 29.

Cuaresma, J. C., Hlouskova, J., Kossmeier, S., Obersteiner, M. (2003): Forecasting Electricity Spot Prices Using Linear Univariate Time Series Models. In: Applied Energy 76 (forthcoming).

Dahl, C., Erdogan, M. (1994): Econometric Energy Demand and Supply Elasticities: Truth or Fiction. In: Proceedings of the 17th Annual International Energy Conference. Cleveland.

Dahlgren, R., Liu, C. - C., Lawarree, J. (2003): Risk Assessment in Energy Trading. In: IEEE Transactions on power systems 18, pp. 503 - 511.

de Jong, C., Huisman, R. (2002): Option Formulas for Mean - Reverting Power Prices with Spikes. Working Paper. Energy Global, Rotterdam School of Management at Erasmus University Rotterdam. Rotterdam, also: http://papers.ssrn.com/sol3/papers.cfm?abstract_id=324520, accessed August 4, 2003.

de Vries, L., Hakvoort, R. (2003): The Question of Generation Adequacy in Liberalized Electricity Markets. In: Proceedings of the 26th Annual IAEE Conference, June 4 – 6, 2003. Prague.

Dennerlein, R.(1990): Energieverbrauch privater Haushalte. Augsburg.

Dentcheva, D.; Römisch, W. (1998): Optimal power generation under uncertainty via stochastic programming. In Marti, K., Kall, P. (eds.): Stochastic Programming Methods and Technical Applications. Lecture Notes in Economics and Mathematical Systems vol. 458. Berlin.

Denton, M., Palmer, A., Masiello, R., Skantze, P. (2003): Managing Market Risk. In Energy. In: IEEE Transactions on power systems 18, pp. 494 - 502.

Deutsche Bundesbank (2003): Emission von Bundesanleihen, Bundesobligationen, Bundesschatzanweisungen und Unverzinslichen Schatzanweisungen des Bundes seit 1999 (chronologisch). http://www.bundesbank.de/kredit/download/embwp.pdf, accessed August 4, 2003.

Diebold, F.X., Schuermann, T., Stroughair, J.D. (1998): Pitfalls and opportunities in the use of extreme value theory in risk management. In: Refenes, A. P.N., Moody, J.D., Burgess A.N. (eds.): Decision Technologies for Computational Finance. Amsterdam, pp. 3 - 12.

Dino, R. N. (1985): Forecasting the Price Evolution of New Electronic Products. In: Journal of Forecasting 4, 1, pp. 39 - 60.

Dixit, A. K., Pindyck, R. S. (1994): Investment under Uncertainty. Princeton.

Dörner, D., Horváth, P., Kagermann, H. (2000): Praxis des Risikomanagements: Grundlagen, Kategorien, branchenspezifische und strukturelle Aspekte. Stuttgart.

Dorris, G., Dunn, A. (2001): Earnings at risk: Better for asset owners. In: Global Energy Business June 2001, pp. 8 – 11.

Dotzauer, E., Lindberg, P. O. (2001): An Application of Lagrangian Relaxation in Short - Term Scheduling of a Cogeneration Plant Under Uncertainty. In: Dotzauer, E: Energy System Operation by Lagrangian Relaxation. Ph. D. Linköping University, Linköping.

Dubin, J. A., McFadden, D. L. (1984): An Econometric Analysis of Residential Electric Appliance Holdings and Consumption. In: Econometrica 52, pp. 345 – 362.

Dutton, J. M., Thomas, A. (1984): Treating Progress Functions as a Managerial Opportunity. In: Academy of Management Review 9, pp. 235 - 247.

DVG – Deutsche Verbundgesellschaft (2000): GridCode 2000 – Netz - und Systemregeln der deutschen Übertragungsnetzbetreiber. Heidelberg.

E.On (2003): Geschäftsbericht 2002. Düsseldorf.

EC – European Commission (2002): Energy in Europe. 2001 – Annual Energy Review. Luxemburg

Economides, N. (1989): Desirability of Compatibility in the Absence of Network Externalities. In: American Economic Review 78, pp. 108 - 121.

EEX (2002): Trading results. http://www.eex.de, accessed August 4, 2003.

EIA – Energy Information Administration (2003): Country Analysis Briefs provided under http://www.eia.doe.gov/emeu/cabs/contents.html, accessed January 7, 2003.

Ellersdorfer, I. (2003): Modellierung von Wettbewerb in der Stromwirtschaft. Ph. D. under preparation. Stuttgart, Oldenburg.

Ellersdorfer, I., Blesl, M., Traber, T, Fahl, U., Kessler, A. (2003): Marktposition deutscher EVU im liberalisierten europäischen Elektrizitätsmarkt. – Analysen mit einem spieltheoretischen Modellansatz. In: Energiewirtschaftliche Tagesfragen 53 (forthcoming).

Ellersdorfer, I., Specht, H., Fahl, U. Voß, A. (2001): Wettbewerb und Energieversorgungs-strukturen der Zukunft. Forschungsbericht des IER, Universität Stuttgart Band 81.

Elliott, R. J., Sick, G. A. Stein, M. (2000): Pricing Electricity Calls. Working paper, University of Alberta. Calgary. http://www.ucalgary.ca/~sick/Research/ElliottSickStein Electric5.pdf, accessed August 4, 2003.

Embrechts, P., McNeil, A., Straumann, D. (2000): Correlation and Dependence in Risk Management: Properties and Pitfalls. In: Dempster, M. A. H. (ed.): Risk Management: Value at Risk and Beyond, Cambridge, pp. 176 - 223.

Embrechts, P., Klüppelberg, C., Mikosch, T. (1997): Modeling Extremal Events for Insurance and Finance. Berlin.

Embrechts, P., Lindskog, F., McNeil, A. (2003): Modelling Dependence with Copulas and Applications to Risk Management. In: Rachev, S. (ed.): Handbook of Heavy Tailed Distributions in Finance, Amsterdam, pp. 329 - 384.

EnBW (2002a): Innovationsbericht 2002. Karlsruhe.

EnBW (2002b): Vorkommerzielles 1 MW - Brennstoffzellen - Kraftwerk in Marbach/Neckar wird nicht gebaut. Pressemitteilung vom 19.12.2002. http://www.enbw.com/content/de /presse/pressemitteilungen/2002/12/12_19_2002__Brennstoffzelle_Marbach_nicht_gebaut _/ index.php, accessed August 4, 2003.

EnBW (2003): Geschäftsbericht 2002. Karlsruhe.

Ender, D. (2003): Windenergienutzung in der Bundesrepublik Deutschland - Stand 31.12.2002. In: DEWI Magazin Nr. 22, Februar 2003, pp. 7 - 19

Engle, R., Kroner, K. (1995): Multivariate simultaneous generalized Arch. In: Econometric Theory 11, pp. 122 – 150.

Engle, R., Lee, G. (1993): A permanent and transitory component model of stock return volatility. UCSD discussion paper 92 - 44R.

Engle, R., Mezrich, J. (1996): Garch for Groups. In: RISK 9, pp. 36 - 40.

Engle, R.F., Ng, V., Rothschild, M. (1990): Asset pricing with a Factor - ARCH covariance structure: Empirical estimates for Treasury bills. In: Journal of Econometrics 45, pp. 213 - 237.

Enquete Commission (2002): Endbericht der Enquete - Kommission „Nachhaltige Energie-versorgung unter den Bedingungen der Globalisierung und der Liberalisierung". Druck-sache des Dt. Bundestages 14/9400. Bonn.

Erdmann, G. (1995): An Evolutionary Model for Long Term Oil Price Forecasts. In: Wagner, A., Lorenz, H. (eds.): Studien zur Evolutorischen Ökonomik III. Berlin, München, pp. 143 - 161.

Erdmann, G., Federico, T. (2001). Time Series Analysis and Forecasting of German Peakload Prices. The Approach of www.prognoseforum.de. Presentation at 2. Internationale Energiewirtschafts-Tagung, 13. –15. Februar 2001. TU Wien.

ETSO – European Transport System Operators (2003): Internet site http://www.etso - net.org; accessed August 4, 2003.

EU – European Union (1997): Directive 96/92/EC of the European Parliament and of the Council of 19 December 1996 concerning common rules for the internal market in electricity. In: Official Journal L 027, January 30, 1997, pp. 20 - 29

Fama, F.F., French, K.R. (1987): Commodity Futures Prices: Some Evidence on Forecast Power, Premiums, and the Theory of Storage. In: The Journal of Business 60, pp. 55 - 73.

Farrukh, C.J.P., Phaal, R. Probert, D.R. (2000): TMAP - A Guide to Technology Management Assessment Procedure. London.

FCCG - Fuel Cell Commercialization Group (2001). Internet site.http://www.ttcorp.com/fccg/index.htm, accessed June 25, 2001.

FERC – Federal Energy Regulatory Commission (2002): Notice of Proposed Rulemaking. Docket No. RM01-12-000. http://www.ferc.gov/Electric/RTO/Mrkt-Strct-comments/nopr/ Web-NOPR.pdf, accessed August 6, 2003.

Ferguson, T.S. (1973): A Bayesian Analysis of Some Nonparametric Problems. In: The Annals of Statistics 2, pp. 615 - 629.

Ferguson, T.S. (1983): Bayesian Density Estimation by Mixtures of Normal Distributions. In Rivisi, H., Rustagi, J. (eds.): Recent Advances in Statistics. New York, pp. 287 - 302.

Fichtner, W., Clemens, C. (2002): Strommodelle - Die Entwicklung des deutschen Strommarktes im europäischen Kontext. Presentation at the workshop: Umwelt - und Klimaschutz in liberalisierten Energiemärkten - Die Rolle erneuerbarer Energieträger, organized by Forum für Energiemodelle, Berlin March 21, 2002. also: http://www.ier.uni - stuttgart.de/public/de/organisation/abt/esa/projekte/forum/index/a_index.htm, accessed August 4, 2003.

Fleten, S.-E., Wallace, S. W., Ziemba, W. T. (2002): Hedging electricity portfolios via stochastic programming. In: Greengard, C., Ruszczynski, A. (eds.): Decision Making Under Uncertainty: Energy and Power. IMA Volumes on Mathematics and Its Applications vol. 128. New York, pp. 71-93.

Fleten, S.-E., Wallace, S. W. (1998): Power Scheduling with Forward Contracts. In: Proceedings of the Nordic MPS, May 9-10, 1998. Molde.

Föllinger, O. (1985): Regelungstechnik – Einführung in die Methoden und ihre Anwendung. Heidelberg.

Fraktionen von CDU/CSU und FDP (2002): Sondervotum der Fraktionen von CDU/CSU und FDP einschließlich der von ihnen benannten Sachverständigen zum Gesamtbericht. In: Endbericht der Enquete - Kommission „Nachhaltige Energieversorgung unter den Bedingungen der Globalisierung und der Liberalisierung". Drucksache des Dt. Bundestages 14/9400, pp. 511 – 624.

Frometa, E., van Aacken, M., Ellmer, M. (2001): Stromgestehungskosten von Großkraftwerken. Paper for the Fachgebiet Energie - und Rohstoffwesen, TU Berlin, Prof. Winje, also: http://www.energiewirtschaft.tu-berlin.de/veranstaltung/files/ ref_stromgestehungskosten.pdf, accessed August 5, 2002.

Fusaro, P. (1998): Energy risk management. New York.

Geweke, J., Amisano, G. (2001): Compound Markov Mixture Models with Applications in Finance. Paper for Econometrics and Statistics Seminar. University of Chicago Graduate School of Business Spring 2001. http://gsbwww.uchicago.edu/fac/robert.mcculloch /research/seminar/2001/GewekeSeminar.pdf, accessed August 4, 2003.

Ghemawat, P. (1985): Building strategy on the experience curve. In: Harvard Business Review, pp. 143 - 149.

Gibbs - Sampling Approaches with Applications. Cambridge Mass.

282

Gibson, R., Schwartz, E. S. (1990): Stochastic Convenience Yield and the Pricing of Oil Contingent Claims. In: The Journal of Finance 45, pp. 959 - 976.

Gilbert, R., Neuhoff, C., Newbery, D. (2002): Mediating Market Power in Networks. In: Proceedings of the 25th Annual IAEE International Conference, Aberdeen June 26 – 29, 2002.

Gilsdorf, K. (1995): Testing for Subadditivity of Vertically-Integrated Electric Utilities. In: Southern Economic Journal 18

Goceliakova, Z. (2003): Real option model for energy and ancillary services markets. IIASA Interim Report IR-03-016. Laxenburg.

Goldemberg, J. (1996): The evolution of ethanol cost in Brazil. In: Energy Policy 24, pp. 1127 - 1128.

Grant, R. M. (2003): Strategic planning in a turbulent environment: Evidence from the oil Majors. In: Strategic Management Journal 24, pp. 491 - 517.

Green, R.J., Newbery, D.M. (1992): Competition in the British electricity spot market. In: Journal of Political Economy 100, pp. 929 - 953.

Greene, W. H. (2000): Econometric analysis. 4th ed. Upper Saddle River NJ.

Grobbel, C. (1999): Competition in Electricity Generation in Germany and Neighboring Countries from a System Dynamics Perspective. Frankfurt.

Group of Thirty (1993): Derivatives: Practices and Principles. Washington.

Gröwe-Kuska, N., Kiwiel, K.C., Nowak, M.P., Römisch, W., Wegner, I.(2002): Power management in a hydro - thermal system under uncertainty by Lagrangian relaxation. In: Greengard, C., Ruszczynski, A. (eds.): Decision Making under Uncertainty: Energy and Power, IMA Volumes in Mathematics and its Applications 128, New York et al., pp. 39 - 70.

GRS - Gesellschaft für Reaktorsicherheit (1989): Deutsche Risikostudie Kernkraftwerke Phase B. Report GRS-A-1600. Köln.

Grunwald, A. (2002): Technikfolgenabschätzung - eine Einführung. Berlin.

Guth, L., Sepetys, K. (2001): Cash flow at risk for nonfinancial companies. In: Global Energy Business June 2001, pp. 12 – 14.

Hamilton, J. D. (1994). Time - series analysis. Princeton NJ, 1994.

Hamilton, J.D. (1989): A new approach to the economic analysis of non - stationary time series and the business cycle. In: Econometrica 57, pp. 357 –3 84.

Hanselmann, M. (1996): Entwicklung eines Programmsystems zur Optimierung der Fahrweise von Kraft - Wärme - Koppelungsanlagen. Forschungsbericht des IER, Universität Stuttgart Band 29. Stuttgart.

Harrison, M., Kreps, D. (1979): Martingales and arbitrage in multiperiod securities markets. In: Journal of Economic Theory 20, pp. 381 - 408.

Hartung, T. (2001): Kritische Betrachtung marktorientierter Kapitalkostenbestimmung bei der Bewertung von Versicherungsunternehmen. Presentation at Jahrestagung des Deutscher Verein für Versicherungswissenschaften. also: http://www.versicherungen - competence - center.de/versicherungen.nsf/B41E28CC147997B4C1256B7A0049637D/$File/kapitalkost en - bestimmung.pdf, accessed August 4, 2003.

Harvey, A.C. (1989): Forecasting, Structural Time Series and the Kalman Filter, Cambridge.

Harvey, S. M., Hogan, W. W. (2000): Nodal and Zonal Congestion Management and the Exercise of Market Power. http://ksghome.harvard.edu/~.whogan.cbg.Ksg /zonal_jan10.pdf, accessed August 6, 2003

Hayek, F. v. (1937): Economics and Knowledge. In: Economica 4, pp. 37 - 54.

Henney, A., Keers, G. (1998): Managing Total Corporate Electricity/Energy Market Risks. In: The Electricity Journal 11 (8), pp. 36 – 46.

HEPG - Harvard Electricity Policy Group (2002): Twenty - Ninth Plenary Session. Rapporteur's summary. September 26 - 27, 2002. http://www.ksg.harvard.edu/hepg /Seminar%20Summaries/0902hepgsumm.pdf, accessed August 4, 2003.

Hobbs, B. F. (2001): Linear Complementarity Models of Nash–Cournot Competition in Bilateral and POOLCO Power Markets. In: IEEE Transactions on Power Systems 16, pp. 194 – 202.

Hobbs, B. F. Metzler, C., Pang J.S. (2000): Calculating Equilibria in Imperfectly Competitive Power Markets: An MPEC Approach. In: IEEE Transactions on Power Systems 15, pp. 638 - 645.

Hobbs, B. F., Rothkopf, M. H., O'Neill, R. P., Chao, H.-P. (eds.) (2001): The Next Generation of Unit Commitment Models, International Series in Operations Research & Management Science. Norwell MA

Hönig, P. (1999): Mehrstufiges Verfahren zur Tageseinsatzoptimierung großer Kraft - Wärme - gekoppelter Energieversorgungssysteme. Dissertation Universität Stuttgart 1999.

Horlock, J. H. (1995): Combined Power Plants – Past, Present, and Future. In: Journal of Engineering for Gas Turbines and Power 177, October. pp. 608 – 616.

Horváth, P., Gleich, R. (2000): Controlling als Teil des Risikomanagements: In: Dörner, D., Horváth, P., Kagermann, H. (eds.): Praxis des Risikomanagements. Stuttgart.

Hufendiek, K. (2001): Systematische Entwicklung von Lastprognosesystemen auf der Basis neuronaler Netze. VDI-Fortschritt-Berichte Reihe 6 Energietechnik Nr. 455. Düsseldorf.

Hufendiek, K. (2002): Organisation des Stromhandels. Presentation on June 11, 2002 within the lecture "Energiemärkte und Energiehandel" by Weber, C. at the University of Stuttgart.

Huisman, R., Mahieu, R. (2001): Regime Jumps in Electricity Prices. Working Paper. Energy Global, Rotterdam School of Management at Erasmus University Rotterdam. Rotterdam, also: http://papers.ssrn.com/paper.taf?abstract_id=271910 accessed August 4, 2003.

Hull, J. C. (2000): Options Futures and Other Derivatives. 4th ed. Upper Saddle River NJ.

Huonker, U. (2003): Ein Werkzeug zur Lagrange-Optimierung in der Kraftwerks-einsatzplanung. VDI-Fortschritt-Berichte Reihe 6 Energietechnik, Band 497. Düsseldorf.

Hsu, M. (1998): Spark Spread Options are Hot!.In: The Electricity Journal 11, pp. 1-12.

IEA – International Energy Agency (2001): Competition in electricity markets. Paris.

IEA – International Energy Agency (2002): World Energy Outlook: 2002. Paris. also: http://www.nea.fr/html/ndd/investment/session1/Birol.pdf, accessed August 4, 2003.

ILEX (2003): Italian wholesale electricity model IGen. http://www.ilex.co.uk/news /sndownload.php?entrynr=88, accessed May 23, 2003.

ILOG (2003): ILOG CPLEX. http://www.ilog.com/products/cplex/, accessed August 4, 2003.

J. P. Morgan & Co. (1994): RiskMetrics™, Technical Document. New York.

Jacobs, J., Freeman, G., Grygier, J., Morton, D., Schultz, G., Staschus, K., Stedinger, J. (1995): SOCRATES - a system for scheduling hydroelectric generation under uncertainty. In: Annals of Operations Research 59, pp. 99 – 133.

Jebjerg, L. Riechmann, C. (2001): Prognose von Anbieterstrategien im Stromgroßhandel – Konzepte und internationale Erfahrungen. In: Proceedings der 2. Internationalen Energiewirtschafts-Tagung 21. - 23. Februar 2001. TU Wien.

Johanning, L., Rudolph, B. (2000): Handbuch Risikomanagement. Bad Soden.

Johnson, B., Barz, G. (1999). Selecting Stochastic Processes for Modelling Electricity Prices. In: Risk Books (ed.): Modelling and the Management of Uncertainty, Haymarket House, London, pp. 3 - 22.

Jopp, K. (2001): Brennstoffzelle drängt mit Tempo auf den Markt. In: VDI nachrichten, 6.7. 2001, p. 9.

Joskow, P. L. (2002): Lessons learned from electricity liberalization in the UK and US - Towards a European Market of Electricity. Presentation at SSPA - Italian Advanced School of Public Administration, June 24, 2002. Rome. also: http://econ - www.mit.edu/faculty/pjoskow/files/JOSKOW-R.pdf, accessed August 4, 2003.

Karesen, K. F., Husby, E. (2002): A Joint State - Space Model for Electricity Spot and Futures Prices. NR Norgsk Regnesentral report No. 965 revised version May 13, 2002, also: http://www.nr.no/documents/samba/research_areas/BFF/JointElMod.pdf, accessed August 4, 2003.

Keitel, R. (2000): Applications with Proton Exchange Membrane (PEM) Fuel Cells for a Deregulated Market Place. In: Brennstoffzellen... effiziente Energietechnik der Zukunft. Tagungsband. Friedrichshafen.

Kim, C.-J., Nelson, C. R. (1999): State - Space Models with Regime Switching: Classical and Gibbs-Sampling Approaches with Application. Cambridge MA.

King, A. J. (2002): Duality and martingales: A stochastic programming perspective on contingent claims. Mathematical Programming B 91, pp. 543 – 562.

Klemperer, P. D., Meyer M. A. (1989): Supply Function Equilibria in Oligopoly under Uncertainty. In: Econometrica 57, pp. 1243 – 1277.

Knieps, G. (2002): Wettbewerb auf den Ferntransportnetzen der deutschen Gaswirtschaft - Eine netzökonomische Analyse. In: Zeitschrift für Energiewirtschaft 26, pp. 171 - 180.

Kolmetz, S., Rouvel, L. (1995): Energieverbrauchsstrukturen im Sektor Haushalte. IKARUS - Instrumente für Klimagas - Reduktionsstrategien, Abschlussbericht Teilprojekt 5 „Haushalte und Kleinverbraucher“, Sektor „Haushalte“. Monographien des Forschungszentrums Jülich Band 17. Jülich.

Kramer, N.: Modellierung von Preisbildungsmechanismen im Strommarkt. Ph. D. Freiberg.

Krasenbrink, B., Nießen, S., Haubrich, H. - J. (1999): Risikomanagement in Stromerzeugung und –handel. In: VDI (ed.): Optimierung in der Energieversorgung II. VDI Berichte 1508, pp. 53 - 62.

Kraus, M. (Hrsg.), (2002): Wetterderivate im Strom - und Gasgroßhandel. Funktionsweise und Einsatz derivativer Finanzinstrumente zur Absicherung von Handelserlösen gegen klimatische Einflüsse. In: Symposium March 11, 2002, Mannheim. http://www.gee.de /old/ws_mar_02.htm, accessed July 14, 2003.

Kreps, D. M., Scheinkman, J. A. (1983): Quantity Precommitment and Bertrand Competition Yield Cournot Outcomes. In: Bell Journal of Economics 14, pp. 326 - 337.

Kreuzberg, M. (1999). Spotpreise und Handelsflüsse auf dem europäischen Strommarkt - Analyse und Simulation. In: Zeitschrift für Energiewirtschaft 23, pp. 43 - 63.

Kühner, R. (1996): Ein verallgemeinertes Schema zur Bildung mathematischer Modelle energiewirtschaftlicher Systeme. Forschungsbericht des IER, Universität Stuttgart Band 35. Stuttgart.

Kuipers, J. (1998): 100 kW SOFC System Experiences From A User Point of View. In: Fuel Cells – Clean Energy for Today's World. Proceedings of the 1998 Fuel Cell Seminar. Palm Springs, November 16 – 19, 1998.

Kurihara, I. et al, (2002): Computer Simulation of Electricity Market - Development of Fundamental Electricity Model using Multi Agent System, CRIEPI Report No T01036, short summary in: CRIEPI, annual research report. http://criepi.denken.or.jp/eng/PR /Nenpo/2002E/02seika19.pdf, accessed July 31, 2003.

KWI (2003): Internet site http://www.kwi.com/, accessed August 4, 2003.

Kydes, A. (2000): Modelling Technology Learning in the National Energy Modelling System. In: Wene, C.-O., Voß, A., Fried, T. (eds.): Experience Curves for Policy Making - The Case of Energy Technologies. Forschungsbericht des IER, Universität Stuttgart Band 67, pp. 181 - 202.

Leuschner, U. (2001): RWE und EnBW gehen an den Kapitalmarkt. http://buerger.metro - polis.de/udo_leuschner/energie-chronik/010414.htm, accessed August 4, 2003.

Leuschner, U. (2002): RWE - Konzern nimmt Anleihe von 6, 4 Milliarden Euro auf. http://buerger.metropolis.de/udo_leuschner/energie-chronik/020403.htm, accessed August 4, 2003.

Lintner, J. (1965): The Valuation of Risk Assets and the Selection of Risky Investments in Stock Portfolios and Capital Budgets. In: Review of Economics and Statistics 47, pp. 13 - 37.

Lipman, T. E., Sperling, D. (2000): Forecasting the Costs of Automotive PEM Fuel Cell Systems – Using Bounded Manufacturing Progress Functions. In: Wene, C.-O., Voß, A., Fried, T. (eds.): Experience Curves for Policy Making - The Case of Energy Technologies. Forschungsbericht des IER, Universität Stuttgart Band 67, pp. 135 - 150.

Longstaff, F. A., Schwartz, E. S. (2001): Valuing American Options by Simulation: A Simple Least-Wquares Approach. In: The Review of Financial Studies 14, pp. 113 – 147.

Luce, R.D., (1959): Individual Choice Behavior: A Theoretical Analysis. New York.

Lucia, J. J., Schwartz, E. S. (2002): Electricity prices and power derivatives: Evidence from the Nordic power market. In: Review of Derivative Research 5, pp. 5 - 50.

Machlup, F. (1942): Competition, Pliopoly and Profit. In: Economica, 9, pp. 1 - 23.

Maddala, G.S. (1992): Limited Dependent and Qualitative Variables in Econometrics. Cambridge.

Madlener, R., Kaufmann, M. (2002): Power exchange spot market trading in Europe: theoretical considerations and empirical evidence. OSCOGEN Deliverable 5.1b. http://www.oscogen.ethz.ch, accessed August 4, 2003.

Margrabe, W. (1978): The Value of an Option to Exchange One Asset for Another. In: Journal of Finance 33, pp.177 - 186.

Mattsson, N., Wene, C.-O. (1997): Assessing new energy technologies using an energy system model with endogenized experience curves. In: International Journal of Energy Research 21, pp. 385 - 393.

McFadden, D. L. (1974): Conditional Logit Analysis of Qualitative Choice Behaviour. In: P. Zarembka (ed.): Frontiers in econometrics (New York), pp. 105 – 142.

McFadden, D. L. (1978): Modelling the Choice of Residential Location. In: A. Karlquist (ed.): Spatial Interaction Theory and Residential Location, Amsterdam, pp. 75 – 96.

Merton, R. C. (1976): Option pricing when underlying stock returns are discontinuous. In: Journal of Financial Economics 3 pp. 125 - 144.

Metzler, C., Hobbs, B.F., Pang J.S. (2003): Nash-Cournot Equilibria in Power Markets on a Linearized DC Network with Arbitrage: Formulations and Properties. In: Networks & Spatial Economics 3, pp. 123 - 150.

Mo, B., Gjelsvik, A., Botterud, A., Grundt, A., Eliasson, B. (1999): Application of a new tool for integrated risk management. In: Proceedings Hydropower into the next century, Gmunden - Austria, pp. 817-826.

Morton, D. P. (1996): An enhanced decomposition algorithm for multistage stochastic hydroelectric scheduling. Annals of Operations Research 64, pp. 211 – 235.

Müsgen, F., Kreuzberg, M. (2001) Einfluss der Speicherkraftwerke in der Alpenregion auf die Strompreise in Deutschland – Eine modellgestützte Analyse. In: Proceedings of 2. Internationale Energiewirtschaftstagung. TU Wien.

N.N. (1998): Gesetz zur Kontrolle und Transparenz im Unternehmensbereich (KonTraG). In: Bundesgesetzblatt Jahrgang 1998 Teil I, pp. 786 – 794.

Nakicenovic, N. (1997): Technological change and learning. In: Perspectives in Energy, volume 4, pp. 173 - 189.

Nelsen R.B. (1999): An Introduction to Copulas. New York.

Nelson, D. B. (1991): Conditional Heteroskedasticity in Asset Returns. In: Econometrica 59, pp. 347– 370.

Nemhauser, G., Wolsey, L. (1988): Integer and Combinatorial Optimization. New York.

Newbery, D. (1998): Competition, Contracts and Entry in the Electricity Spot Market. In: Rand Journal of Economics 29, pp. 726 – 749.

Nitsch, J. (1998): Probleme der Langfristkostenschätzung – Beispiel Regenerative Energien. Presentation at the Workshop „Energiesparen – Klimaschutz, der sich rechnet", October 8-9, 1998. Rottenburg an der Fulda.

Nordpool (2003): Weekly System Prices. http://www.nordpool.no/marketinfo/index.html, accessed August 4, 2003.

OFGEM – Office of Gas and Electricity Markets (2002): The review of the first year of NETA - A review document. http://www.ofgem.gov.uk/temp/ofgem/cache/cmsattach/1984_48neta_year_review.pdf, accessed August 6, 2003.

Ohl, M. (2002): Technische und ökonomische Analyse von Schmelzkarbonatbrennstoffzellen. IER, Universität Stuttgart Diplomarbeit Band 370. Stuttgart.

Oren, S. S. (2000): Capacity Payments and Supply Adequacy in Competitive Electricity Markets. Paper for the VII Symposium of Specialists in Electric Operational and Expansion Planning, May 21 – 26, 2000. Curtiba. also: http://www.ieor.berkeley.edu/%7Eoren/workingp /sepope.pdf, accessed August 4, 2003.

Oren, S., Gross, G., Alvarado, F. (2002): Alternative Business Models for Transmission Investment and Operation. Report to the Department of Energy for the National Transmission Grid Study. Washington.

Ostertag, K., Llerena, P., Richard, A. (eds.) (2004): Option valuation for energy issues. Stuttgart (forthcoming).

Pearce, D. W. (1998): Economics and environment. Essays on ecological economics and sustainable development. Cheltenham.

Pereira, M. V. F., McCoy, M. F., Merrill, H. M. (2000): Managing Risk in the New Power Business. In: IEEE Computer Applications in Power 13 (2), pp. 18 - 24.

Pereira, M. V. F., Pinto, L. M. (1991): Multi - stage stochastic optimization applied to energy planning. In: Mathematical programming A 52, pp. 359 – 375.

Pfaffenberger, W. (2002): Energiepolitische Rahmenbedingungen und Investitionen im Kraftwerksbereich bis 2020. In: Energiewirtschaftliche Tagesfragen 52, pp. 602 – 607.

Pilipovic, D. (1998): Energy Risk. New York et al..

Pindyck, R.S. (2001): The dynamics of commodity spot and futures markets: a primer. In: The Energy Journal 22, pp. 1 - 29.

Pineau, P. - O. (2000): Electricity Market Reforms: Institutional Developments, Investment Dynamics and Game Modeling. PhD thesis. Montreal.

Pohlmann, M., Pospischill, H. (1999): Nur analytisch ohne wirtschaftliche Risiken. In: Zeitschrift für Kommunalwirtschaft (ZfK), p. 8.

Pollitt, M., (1997): The Impact of Liberalization on the Performance of the Electricity Supply Industry: An International Survey. In: Journal of Energy Literature 3, pp. 3 - 31.

Press, W. H., Flannery, B. P., Teukolsky, S. A., Vetterling, W. T. (1988): Numerical Recipes in C: The Art of Scientific Computing. Cambridge University Press.

Prospex (2003): Links provided under http://www.prospex.co.uk/links.htm, accessed January 7, 2003.

Ramanathan, R., Engle, R., Granger, C.W.J., Vahid-Araghi, F., Brace, C. (1997): Short-run forecasts of electricity loads and peaks. In: International Journal of Forecasting, 13, pp. 161 - 174.

Rockafellar, R.T., Uryasev, S. (2000): Optimization of Conditional Value-At-Risk. In: The Journal of Risk 2 (3), pp. 21 - 41.

Römisch, W. (2001): Optimierungsmethoden für die Energiewirtschaft. Stand und Entwicklungstendenzen. VDI (ed.): Optimierung in der Energieversorgung IV. VDI-Berichte Band 1627, Düsseldorf 2001, pp. 23 - 36.

Ross, S. A. (1977): The Capital Asset Pricing Model (CAPM), Short - Sale Restrictions and Related Issues. In: Journal of Finance 32, pp. 177 – 183.

Rudolph, B. (2002): Vorlesung Risikomanagement und Derivate Finanzinstrumente, Kapitel 1. Course material. http://www.kmf.bwl.uni-muenchen.de, accessed July 2, 2002.

Rudolph, B., Johanning, L. (2000): Entwicklungslinien im Risikomanagement. In: Johanning, L. Rudolph, B. (eds.): Handbuch Risikomanagement, vol. 1. Bad Soden, pp. 15 - 52.

Rüffler, W. (2001): Integrierte Ressourcenplanung für Baden - Württemberg. Forschungsbericht des IER, Universität Stuttgart Band 77. Stuttgart.

RWE (2003): Geschäftsbericht 2002. Essen.

Sakamoto, T., Ishiguro, M., Kitagawa, G. (1986): Akaike Information Criterion Statistics. D. Reidel. Holland.

Sander, K. Weber, C. (2001): Stationäre Brennstoffzellen – Wo geht die Reise hin? In: VDI (ed.): Fortschrittliche Energiewandlung und –anwendung, Tagung, Bochum, 13. – 14.03.2001. VDI - Berichte 1594. Düsseldorf.

Sander, K., Weber, C., Blesl, M., Voß, A. (2003): Perspektiven stationärer Brennstoffzellen im Energiesystem Baden - Württembergs. Forschungsbericht FZKA – BWPLUS. http://bwplus.fzk.de/berichte/SBer/BWE20008SBer.pdf accessed December 15, 2003

Savage, L. J. (1954): The Foundations of Statistics. New York.

Schnell, U.; Sauer, Ch.; Hein, K.R.G.; Moser, P.; Winderlich, W. (2002): Detaillierte gekoppelte Prozeßsimulation eines Großkraftwerks (Kurzfassung KOMET - Teilprojekt 2.9). In: VDI (ed.): Modellierung und Simulation von Dampferzeugern und Feuerungen. VDI - Berichte Nr. 1664, pp. 73 - 83.

Schuler, A. (2000): Entwicklung eines Modells zur Analyse des Endenergieeinsatzes in Baden - Württemberg. Forschungsbericht des IER, Universität Stuttgart Band 66. Stuttgart.

Schultz, R., Tiedemann, S. (2002): Risk Aversion via Excess Probabilities in Stochastic Programs with Mixed - Integer Recourse. Schriftenreihe des Instituts für Mathematik SM - DU - 531. Duisburg.

Schwartz, E. (1997): The Stochastic Behavior of Commodity Prices: Implications for Valuation and Hedging. In: The Journal of Finance 52, pp. 923 - 973.

Sen, S., Kothari D. P. (1998): Optimal thermal generating unit commitment – a review. In: Electrical Power & Energy Systems 20, pp. 443 – 451.

Sharma, C. et al. (2002): Business models for Transmission Operation. http://www.stoft.com /e/lib/papers/Sharma - 2002 - transmission - organization.pdf, accessed August 4, 2003.

Sharpe, W. (1964): Capital Asset Prices: A theory of Market Equilibrium under Conditions of Risk. In: Journal of Finance 19, pp. 425 – 442.

Sheble, G., Fahd, G. (1994): Unit commitment literature synopsis. In: IEEE Transactions on power systems 9, pp. 128 – 135.

Shoven, J. B., Whalley, J. (1992): Applying General Equilibrium. Cambridge.

Shuttleworth, G., Lieb-Doczy, E. (2000): Wirtschaftliche Effizienz und wettbewerbliche Aspekte der Bereitstellung von Regelenergie in Deutschland. Ein Gutachten für den Verband kommunaler Unternehmen (VKU), angefertigt von NERA. London

Simpson, M. (1997): Weather hedging – A new alternative in energy risk management. In: Energy & Power Risk Management 2, pp. 4 - 5.

Skantze, P., Gubina, A., Ilic, M. (2000): Bid - based Stochastic Model for Electricity Prices: The Impact of Fundamental Drivers on Market Dynamics. MIT Energy Laboratory Publication MIT EL 00 - 004. Cambridge MA.

Sklar, A. (1996): Random variables, distribution functions, and copulas - a personal look backward and forward. In: Rüschendorff, L., Schweizer, B., Taylor, M. (eds.): Distributions with Fixed Marginals and Related Topics. Institute of Mathematical Statistics, Hayward, CA, pp. 1 – 14.

Smeers, Y., (1997): Computable Equilibrium Models and the Restructuring of the European Electricity and Gas Markets. In: The Energy Journal 18, pp. 1 – 31.

Smith, A. (1962): The wealth of Nations, vol. 2. Seligman, H. A. (ed.). London, New York.

Smith, R. L. (1990): Extreme value theory. In: Ledermann, W. (ed.): Handbook of Applicable Mathematics supplement. Chichester, pp. 437–471.

Snoussi, H., Mohammad - Djafari, A. (2001): Penalized maximum likelihood for multivariate Gaussian mixture. In: Fry, R. - L. (ed.): Bayesian Inference and Maximum Entropy Methods. Proceedings of the MaxEnt Workshops August 2001, pp 36 - 46.

Söderlind, P. (2003): Lecture Notes in Empirical Asset Pricing. http://web.hhs.se/personal/Psoderlind/, accessed August 4, 2003.

Statistisches Bundesamt (2003): Preisindizes Deutschland. Originalwert – Veränderung zum Vorjahresmonat in %. http://www.destatis.de/indicators/d/pre110jd.htm, accessed August 4, 2003.

Stern, B. (1999): TrendProfile Stromversorgung. Hamburg.

Stern, B., Haubrich, H. - J., Ewert, A. (2001): Stochastische Optimierung von Kraftwerkseinsatz und Stromhandel zur Berücksichtigung von Planungsunsicherheiten In: VDI - Berichte, Band 1627: Optimierung in der Energieversorgung, pp. 37 - 49.

Sutcliffe, K. M., Zaheer, A. (1998): Uncertainty in the transaction environment: an empirical test. In: Strategic Management Journal 19, pp. 1 - 23.

Swider, D. (2004): Handelsentscheidungen bei Risiko an Day-ahead-Auktionsmärkten für Spot- und Regelenergie. Ph. D. under preparation. Stuttgart.

Swider, D., Weber, C. (2003a): Ausgestaltung des deutschen Regelenergiemarktes. In: Energiewirtschaftliche Tagesfragen 53, pp. 448 – 453.

Swider, D., Weber, C. (2003b): Gewinnmaximale Angebotserstellung an Auktionsmärkten für Regelenergie. In: VDI (eds.): Optimierung in der Energieversorgung V. VDI-Berichte 1792. Düsseldorf, pp. 173 - 185.

Takriti, S., Krasenbrink, B., Wu, L. S. - Y. (2000): Incorporating fuel constraints and electricity spot prices into the stochastic unit commitment problem. In: Operations Research 48, pp. 268 – 280.

Thorin, E., Brand, H., Weber, C. (2001): Summary of specified general model for CHP system. OSCOGEN Deliverable D1.4. http:/www.oscogen.ethz.ch accessed August 4, 2003.

Thorin, E., Brand, H., Weber, C. (2003): Long - term Optimization of Cogeneration Systems in a Competitive Market Environment. In: Applied Energy (accepted for publication).

Tseng, C. L., Barz, G. (2002): Short - term Generation Asset Valuation: a Real Options Approach. In: Operations Research 50, pp 297 - 310.

Tsuchiya, H. (2000): Learning Curve Cost Analysis for Model Building of Renewable Energy in Japan. In: Wene, C.-O., Voß, A., Fried, T. (eds.): Experience Curves for Policy Making - The Case of Energy Technologies. Forschungsbericht des IER, Universität Stuttgart Band 67, pp. 67 - 76.

UCPTE – Union de Coordination de la Production et du Transport d'Electricité (1998): Spielregel zur primären und sekundären Wirkleistungsregelung in der UCPTE.

University Bielefeld (2001): Internet site department of chemistry, section DC II. http://dc2.uni - bielefeld.de/dc2/fc/folien/brz_25.htm, accessed June 25, 2001.

Uryasev, S (2000): Conditional Value - at - Risk: Optimization Algorithms and Applications. In: Financial Engineering News 14 (2), pp. 1 - 5.

Varian, H. R. (1992): Microeconomic analysis. 3rd edition. New York.

Vattenfall Europe (2003): Geschäftsbericht 2002. Hamburg.

VDEW – Verband der Elektrizitätswirtschaft (1993): Analyse und Prognose des Stromverbrauchs der privaten Haushalte. Frankfurt/Main.

VDEW – Verband der Elektrizitätswirtschaft (1995): Ermittlung der Lastganglinien bei der Benutzung elektrischer Energie durch die bundesdeutschen Haushalte während eines Jahres. VDEW Publikationen. Frankfurt/Main.

VDEW – Verband der Elektrizitätswirtschaft (2000a): Leistung und Arbeit. VDEW Statistik 1998. Frankfurt.

VDEW – Verband der Elektrizitätswirtschaft (2000b): Kundenzufriedenheit bei Haushaltskunden. Ergebnisbericht des VDEW - Kundenfokus. Frankfurt, Heidelberg.

VDEW – Verband der Elektrizitätswirtschaft (2002): Stromverbrauch konstant. Pressemitteilung vom 11. Februar 2002. Frankfurt.

VGB – Technische Vereinigung der Großkraftwerksbetreiber (1998): Verfügbarkeit von Wärmekraftwerken 1988 – 1997. VGB Technisch-wissenschaftliche Berichte VGB-TW 103. Essen.

Visscher, M.L. (1973): Welfare - Maximizing Price and Output with Stochastic Demand: Comment. In: American Economic Review 63, pp. 224 - 229.

von der Fehr, N.-H. M., Harbord, D. C. (1997): Capacity Investment and Competition in Decentralized Electricity Markets. Department of Economics, University of Oslo Memorandum No. 27/97. Oslo.

Voß, A. (2000): Wettbewerbsfähigkeit der verschiedenen Stromerzeugungsarten im liberalisierten Markt. Presentation at the conference „Kraftwerke 2000" organized by VGB, October 10 – 12, 2000. Düsseldorf. also: http://elib.uni-stuttgart.de/opus/volltexte/2000/691/pdf /Vortrag_ VGB3.pdf, accessed August 4, 2003.

Voß, A. (2002a): Aktuelle Situation und Trends im Europäischen und Deutschen Kraftwerksmarkt. Presentation at the conference „Wettbewerbsfähigkeit der Energieerzeugung". Wuppertal, September 18, 2002. also: http://www.ier.uni-stuttgart.de /public/de/publikationen/VortragWuppertal/VortragWuppertal18092002.pdf, accessed August 4, 2003.

Voß, A. (2002b): The Ability of the Various Types of Power Generation to Compete on the Liberalized Market. In: VGB PowerTech 81 (4), pp. 27 – 31.

Voß, A., Kramer, N. (2000): Risikomanagement in der Energiewirtschaft. In: Dörner, D., Horváth, P., Kagermann, H. (eds.): Praxis des Risikomanagements. Stuttgart, pp. 569 – 588.

Wallace, S. W., Fleten, S. - E. (2002): Stochastic programming models in energy. In: Ruszcynski, A., Shapiro, A. (eds.): Stochastic Programming. Handbooks in Operations Research and Management Science 10. Amsterdam.

Walther, W. F. (2000): Risikomanagement im derivativen Geschäft. In: Johanning, L. Rudolph, B. (eds.): Handbuch Risikomanagement, vol. 2. Bad Soden, pp. 701 - 728.

Weber, C. (1999): Konsumentenverhalten und Umwelt. Frankfurt/Main.

Weber, C. (2001): Preiswettbewerb im Strommarkt – zur Bedeutung von Marktmacht, Produktdifferenzierung und Marktsegmentierung. In: Proceedings of 2. Internationale Energiewirtschaftstagung February 21 – 23, 2001, Wien.

Weber, C. (2003): Combining bottom - up and finance modelling for electricity markets. In: Leopold - Wildburger, Rendl, U. F., Wäscher, G. (2003): Operations Research Proceedings 2002 · Selected Papers of the International Conference on Operations Research (SOR 2002), Klagenfurt, September 2 - 5, 2002. Springer: Berlin u. a. 2003, pp. 272 - 277.

Weber, C., Hoeck, C. (2003): Methoden für das Risikomanagement kombinierter Erzeugungs- und Handelsportfolios. Presentation at the 3. Internationale Energiewirtschaftstagung an der TU Wien, February 12 – 14, 2003. Wien.

Weber, C., Sander, K., Neises, C. (2001): Nutzung von Lastprofilen für die Marktsegmentierung. In: Zeitschrift für Energiewirtschaft 25, pp. 89 - 106.

Weber, C., Thorin, E., Hoffmann, A. (2003): Einsatzplanung und Portfolio - Management für KWK - Systeme unter Unsicherheit. In: VDI (eds.): Fortschrittliche Energiewandlung und –anwendung. VDI - Berichte Band 1746, pp. 733 – 742.

Weber, C., Voß, A. (2001): Bewertung von Kraftwerksinvestitionen im liberalisierten Markt. In: Optimierung in der Energieversorgung. VDI-Berichte Band 1627. Düsseldorf , pp. 11 – 21

Wene, C.-O. (2000): Experience Curves: Measuring the Performance of the Black Box. In: Wene, C.-O., Voß, A., Fried, T. (eds.): Experience Curves for Policy Making - The Case of Energy Technologies. Forschungsbericht des IER, Universität Stuttgart Band 67, pp. 53 - 66.

Whitaker, R. (1998): Investment in volume building: the 'virtuous cycle' in PAFC. In: Journal of Power Sources 71, pp. 71 - 74.

Willis, K.G., Garrod, G.D. (1997): Electric Supply Reliability, Estimating the Value of Lost Load. In: Energy Policy 25, pp. 97 – 103.

Wirtschaftsministerium Baden - Württemberg (2002): Energiebericht 2001. Stuttgart.

Witt, J.; Kaltschmitt, M. (2003): Weltweite Nutzung regenerativer Energien. In: BWK 55, pp. 64 – 71.

Wright, T. P. (1936): Factors Affecting the Costs of Airplanes. In: Journal of Aeronautical Sciences 3, pp. 122 - 128.

Yelle, L. E. (1983): Adding Life Cycles to Learning Curves. In: Long Range Planning 16, pp. 82 - 87.

Index

Early Titles in the
INTERNATIONAL SERIES IN
OPERATIONS RESEARCH & MANAGEMENT SCIENCE
Frederick S. Hillier, Series Editor, *Stanford University*

Early Titles in the
INTERNATIONAL SERIES IN
OPERATIONS RESEARCH & MANAGEMENT SCIENCE
(Continued)

Cox, Louis Anthony, Jr. / *RISK ANALYSIS: Foundations, Models and Methods*

Dror, M., L'Ecuyer, P. & Szidarovszky, F. / *MODELING UNCERTAINTY: An Examination of Stochastic Theory, Methods, and Applications*

Dokuchaev, N. / *DYNAMIC PORTFOLIO STRATEGIES: Quantitative Methods and Empirical Rules for Incomplete Information*

Sarker, R., Mohammadian, M. & Yao, X. / *EVOLUTIONARY OPTIMIZATION*

Demeulemeester, R. & Herroelen, W. / *PROJECT SCHEDULING: A Research Handbook*

Gazis, D.C. / *TRAFFIC THEORY*

* A list of the more recent publications in the series is at the front of the book *